MOON-TO-MARS ARCHITECTURE DEFINITION DOCUMENT (ESDMD-001)

NASA. EXPLORATION SYSTEMS DEVELOPMENT MISSION DIRECTORATE

NIMBLE BOOKS LLC: THE AI LAB FOR BOOK-LOVERS

~ FRED ZIMMERMAN, EDITOR ~

Humans and AI making books richer, more diverse, and more surprising.

PUBLISHING INFORMATION

(c) 2024 Nimble Books LLC
ISBN: 978-1-60888-285-4

AI-GENERATED KEYWORD PHRASES

Moon-to-Mars Architecture Definition Document; NASA exploration; Moon and Mars; high-level structure; framework for exploration; decomposition of objectives; characteristics, needs, use cases, and functions; human systems and habitation; interoperability and standardization; strategic assessments; key considerations; lunar architecture pillars; science, inspiration, and national posture; Human Lunar Return segment; Foundational Exploration segment; Sustained Lunar Evolution segment; Humans to Mars segment; objectives, use cases, and functions.

PUBLISHER'S NOTES

The great thing about this document is that it exists: humans are serious returning to the Moon, and then to Mars. This is the US plan, but it is not clear yet whether it will be China or the US that wins the undeclared but fierce race back to the Moon. In the words of Mike Tyson, "Everyone has a plan until they get punched in the mouth."

Readers should want to read this document because it provides a comprehensive and detailed overview of NASA's Moon-to-Mars architecture, offering insights into the future of space exploration and the potential for scientific and technological advancements that will benefit humanity. It addresses current public issues such as the need for sustainable space exploration, the importance of scientific research and discovery, and the potential for economic opportunities in space. By understanding the architecture and its goals, readers can gain insights into how NASA is planning to proceed, and what the obstacles may be.

This annotated edition illustrates the capabilities of the AI Lab for Book-Lovers to add context and ease-of-use to manuscripts. It includes several types of abstracts, building from simplest to more complex: TLDR (one word), ELI5, TLDR (vanilla), Scientific Style, and Action Items; essays to increase viewpoint diversity, such as Grounds for Dissent, Red Team Critique, and MAGA Perspective; and Notable Passages and Nutshell Summaries for each page.

ANNOTATIONS

ABSTRACTS

TL;DR (ONE WORD)

Comprehensive.

EXPLAIN IT TO ME LIKE I'M FIVE YEARS OLD

This document is like a big plan for NASA's exploration of the Moon and Mars. It tells us what we want to do and how we're going to do it. It talks about things like where people will live and how they will work in space. It also says we need to make sure everything can work together and follow the same rules. The document talks about different parts of the plan, like going back to the Moon and then going to Mars. It explains what we want to learn and how

TL;DR (VANILLA)

This document provides an overview of NASA's exploration plans for the Moon and Mars. It outlines the structure and objectives of the architecture, emphasizing human systems and interoperability. It discusses strategic assessments and key considerations, as well as different segments of the exploration campaign. The document provides information on objectives, use cases, and functions associated with each segment.

SCIENTIFIC STYLE

This document is an Architecture Definition Document that outlines NASA's exploration plans for the Moon and Mars. It provides a high-level overview of the architecture and objectives, including the decomposition of objectives into characteristics, needs, use cases, and functions. The document emphasizes the importance of human systems and habitation, as well as interoperability and standardization across the architecture. It discusses strategic assessments and key considerations for the lunar architecture, highlighting the pillars of science, inspiration, and national posture. The document also provides information on different segments of the exploration campaign, such as Human Lunar Return, Foundational Exploration, Sustained Lunar Evolution, and Humans to Mars segments. It includes objectives, use cases, and functions associated with each segment.

Overall, this document presents a comprehensive overview of the Moon-to-Mars architecture and its components while highlighting the goals and capabilities of each segment.

ACTION ITEMS

Review the Moon-to-Mars Architecture Definition Document to gain a thorough understanding of NASA's exploration plans and objectives.

Identify any areas of interest or relevance to your own work or research.

Consider how your organization or field can contribute to or collaborate with NASA's exploration efforts.

Explore opportunities for interoperability and standardization within your own architecture or system design.

Stay updated on any updates or revisions to the Moon-to-Mars architecture as NASA continues its exploration efforts.

VIEWPOINTS

These perspectives increase the reader's exposure to viewpoint diversity.

GROUNDS FOR DISSENT

A member of the organization responsible for this document might have principled, substantive reasons to dissent from this report due to various factors. Here are a few possible dissenting views:

Financial Concerns: One possible reason for dissent could be related to the financial implications of the Moon-to-Mars architecture. This member might argue that the proposed exploration campaign is too expensive and that NASA's resources would be better allocated towards other pressing priorities, such as climate change research or improving Earth-based technologies.

Ethical Considerations: Another potential reason for dissent could be rooted in ethical concerns. The member might argue that sending humans to explore and potentially colonize other celestial bodies raises significant ethical questions regarding the rights of extraterrestrial life forms, issues of territorial claims, or the potential destruction of unique environments on the Moon or Mars.

Scientific Priorities: Dissent may arise from a scientific standpoint, where a member believes that the Moon-to-Mars architecture places undue emphasis on human systems and habitation at the expense of scientific research. They might contend that more robotic missions should be prioritized to gather data and conduct experiments without risking human lives.

Alternative Approaches: A dissenter might believe that there are alternative approaches or destinations within space exploration that warrant greater focus instead of the Moon and Mars. They could argue for redirecting resources towards exploring other celestial bodies like Europa or Enceladus, which show promising signs of potential extraterrestrial life.

International Collaboration: The member might take issue with the document's emphasis on national posture and propose a more collaborative approach involving international partners. They may argue that fostering cooperation with other space-faring nations would lead to shared costs, increased scientific discoveries, and a more inclusive exploration effort.

Safety Concerns: Dissenting views could also stem from concerns over safety risks associated with human space exploration beyond low-Earth orbit. The member might argue that current technologies and understanding of long-duration space travel and habitation are insufficient to ensure the safety and well-being of astronauts during extended missions to the Moon or Mars.

Impact on Earth: Lastly, a member might dissent based on the potential negative consequences that could arise from diverting significant resources towards Moon-to-Mars exploration. They might argue that these resources could be better utilized to address pressing issues on Earth, such as poverty alleviation, healthcare, or environmental conservation.

These dissenting views highlight possible principled and substantive reasons why a member of the organization responsible for this document might choose to disagree with its contents. It is important to note that these arguments are hypothetical and may not reflect the actual opinions of individuals within the organization.

RED TEAM CRITIQUE

The Moon-to-Mars Architecture Definition Document provides a comprehensive overview of NASA's exploration plans for the Moon and Mars. It effectively outlines the high-level structure and framework for the exploration, ensuring that all objectives are decomposed into characteristics, needs, use cases, and functions. This level of detail is crucial for understanding the overall architecture.

The document appropriately emphasizes the importance of human systems and habitation in space exploration. By acknowledging this key aspect, NASA shows its commitment to ensuring astronaut safety and well-being during extended missions to celestial bodies. This focus on human systems aligns with best practices in space exploration.

Interoperability and standardization are rightly identified as key considerations across the architecture. These factors are crucial for effective collaboration between different components of the mission. By establishing interoperability standards early on, NASA can minimize potential issues or bottlenecks that may arise during later stages of implementation.

Strategic assessments and key considerations regarding lunar architecture showcase a thoughtful approach to goal-setting within this context. The pillars of science, inspiration, and national posture demonstrate an understanding of various stakeholders' interests while ensuring scientific advancements remain at the core of these missions.

Segmenting the exploration campaign into Human Lunar Return, Foundational Exploration, Sustained Lunar Evolution, and Humans to Mars segments helps provide a coherent structure to understand different phases of activity along with their associated objectives, use cases,and functions.This segmentation allows stakeholders to grasp each segment's unique goals more clearly by breaking them down into manageable parts.

The inclusion of specific information about each segment further enhances understanding by providing concrete details about what is expected from these individual phases.This specificity enables readers to evaluate how well each phase contributes toward achieving ultimate objectives.

Overall,the document successfully presents a cohesive pictureoftheMoon-to-Mars architecture and its components. It provides a clear roadmap for NASA's exploration plans and adequately addresses key considerations such as human systems and interoperability/standard-ization. The strategic assessments and segment breakdowns add depth, ensuring further understanding and transparency. This document provides a strong foundation for future planning and implementation of Moon-to-Mars exploration endeavors.

MAGA PERSPECTIVE

This document is just another example of NASA wasting taxpayer dollars on pointless space exploration. Instead of focusing on the needs and priorities of the American people here on Earth, they continue to push for missions to the Moon and Mars that serve no purpose other than to satisfy the egos of scientists and bureaucrats.

The emphasis on human systems and habitation is misguided. We should be investing in our own infrastructure, healthcare, and education before wasting resources on building habitats on distant celestial bodies. The focus should be on improving the lives of Americans, not sending a handful of astronauts to live in isolation in space.

Interoperability and standardization across the architecture? That just sounds like more government bureaucracy and red tape. This kind of thinking is what holds back innovation and progress. Instead of stifling creativity with rules and regulations, we should be encouraging private companies to lead the way in space exploration.

And let's not forget about the strategic assessments and key considerations for the lunar architecture. Science, inspiration, and national posture? These are nothing more than empty buzzwords used to justify an unnecessary mission. We don't need inspiration or national posture; we need practical solutions to real problems facing our nation.

Finally, the segmentation of the exploration campaign into different segments seems overly complicated and convoluted. Why can't NASA simply focus on one goal at a time instead of spreading its resources thin across multiple objectives? This lack of prioritization only further demonstrates the inefficiency and wastefulness of our space program.

PAGE-BY-PAGE SUMMARIES

features into representative reference missions and concepts of operations, as well as the need for an iterative framework process to accommodate the scale of the architecture.

stakeholder needs and long-term goals. The Responsible Use Tenet ensures consistent application of policy and ethical frameworks.

BODY-27 *The page discusses the foundational capabilities needed for lunar and Mars architectural decisions, including support systems for crew operations, sample retrieval, technology demonstration, and resource utilization. It also highlights the importance of testing environments and the need for two key destination platforms: long-duration microgravity systems and surface platforms for partial gravity. Additionally, it mentions leveraging low-earth-orbit assets and optimizing system access to the lunar South Pole.*

BODY-28 *The lunar South Pole is a key location for the Moon-to-Mars Architecture due to its performance drivers, scientific opportunities, and enabling characteristics. The lighting conditions at the South Pole provide flexibility for global locations and longer-duration operations, while also preserving volatiles and offering valuable resources for future exploration. The South Pole's unique features make it an important part of the integrated architecture.*

BODY-29 *The Moon-to-Mars architecture will utilize the Near Rectilinear Halo Orbit (NRHO) for lunar microgravity staging operations. Crew transportation systems will be developed to ensure safety and contingency planning. The ability to transport crew, cargo, and support systems efficiently and safely is crucial for achieving mission objectives. Mobility systems on the lunar surface are necessary for maximizing crew time and enabling exploration in inaccessible areas. A robust communication system is also critical for monitoring and controlling vehicles.*

BODY-30 *The Moon-to-Mars architecture aims to handle multiple data streams and support annual crewed lunar missions. It emphasizes the need for logistics supply, repurpose, and disposal, as well as commercial partnerships for sustainability. The architecture includes systems for up to four crew members to conduct scientific and technological objectives.*

BODY-31 *The Moon-to-Mars architecture focuses on the safety and reliability of systems for human exploration missions. Human-rating is crucial, incorporating design features to enhance safety and crew recovery. Collaboration with industry and international partners is essential for innovation and economic development. NASA aims to broaden global partnerships and demonstrate operational flexibility for long-term Mars-forward development.*

BODY-32 *The document discusses the unique considerations for lunar exploration and the strategic assessments for Mars architecture. It emphasizes the need for flexibility, incremental increase in capabilities, and addressing stakeholder needs in order to build a successful architecture.*

BODY-33 *The page discusses the importance of the timing decision in planning a Mars exploration architecture, using the Apollo program as an example. It highlights how different decisions are influenced by the chosen timeline and emphasizes the need for considering technological development and mission duration.*

BODY-34 *NASA's Moon-to-Mars architecture development approach focuses on the objectives and reasons for deep space exploration before determining when and how to proceed. Prioritizing different objectives will influence decisions on landing sites, mission elements, and crew selection.*

BODY-35 *The Mars architecture decision flow is iterative and dependent on prioritized decisions. The distance between Earth and Mars presents unique considerations, as it is far beyond previous human space flight experiences. Operational experience and paradigms will require a different approach than heritage programs.*

a higher propulsion mass, while longer missions benefit from favorable alignments between Earth and Mars. The decision on Mars vicinity stay time depends on surface mission duration, orbital requirements, and total roundtrip mission duration. The mission operation mode assumes an "all-up" mode with pre-deployed surface assets.

mentions ongoing assessments and future work related to infrastructure needs, trade studies, and human health research for Mars missions.

BODY-111   The page discusses the assessment of recurring tenets in NASA's Moon-to-Mars architecture, focusing on international collaboration. It highlights existing partnerships and potential contributions from international partners such as the European Space Agency, Japan Aerospace Exploration Agency, and Canadian Space Agency.

BODY-112   NASA is actively engaging in international partnerships for lunar science, space communications, additional capabilities, and technology development. These collaborations aim to advance science, exploration, and space technology goals while also promoting education and public engagement. The Moon-to-Mars architecture will identify potential gaps and opportunities for cooperation.

BODY-113   NASA engages in international discussions and partnerships for Moon-to-Mars exploration. Multilateral forums, such as ISECG and LEAG, are used to articulate objectives and identify areas of cooperation. NASA also collaborates with industry to achieve common goals.

BODY-114   The Moon-to-Mars architecture relies on partnerships with U.S. industry for exploration services. However, there are gaps in capturing and leveraging industry contributions. The safety of the crew is a top priority, but there are knowledge gaps regarding the effects of long-term exposure to deep space and contingency scenarios. Maximizing crew time for scientific activities is also important.

BODY-115   The Moon-to-Mars architecture aims to maximize crew time for scientific activities and minimize maintenance and construction tasks. Knowledge gaps exist regarding infrastructure and capabilities, requiring further assessment for system design and operational planning. Reusability and maintainability are important for long-term sustainability, but their risks and impacts need to be understood. Concerns remain about asset lifespan, maintenance requirements, refurbishment, and system limitations.

BODY-116   The Moon-to-Mars architecture is being developed to adhere to existing laws and policies, including planetary protection. However, there are policy gaps that need to be addressed. Interoperability is a priority, with baseline interfaces being identified but specific implementations still in development. Standardization policies are needed for cross-program and cross-partner element development.

BODY-117   The Moon-to-Mars architecture aims to leverage infrastructure in low-Earth orbit and foster the expansion of commerce and space development. However, further studies and refinement are needed to explore all available options and address gaps in current plans for commercialization beyond LEO.

BODY-118   This page is an appendix that provides a comprehensive breakdown of lunar objectives related to science, including functions, use cases, and characteristics. It also highlights specific segments related to human lunar return missions.

BODY-119   This page outlines various functions, use cases, and objectives related to the Moon-to-Mars architecture. It includes tasks such as recovering and packaging surface samples, conducting crew surveys, transporting cargo, and providing power for deployed payloads. The goal is to advance understanding of geologic processes on planetary bodies.

BODY-120   This page outlines various functions and use cases related to the Moon-to-Mars architecture, including high bandwidth communication, sample collection and storage, surface activities, and cargo transport.

NOTABLE PASSAGES

BODY-4   *"To enable this effort, long-term goals and objectives have been established in the Moon-to-Mars Strategy and Objectives document; however, the practical management and execution to ensure objective satisfaction requires an innovative approach to the definition of NASA's Moon-to-Mars human exploration architecture. Architecture is the high-level unifying structure that defines a system. It provides a set of rules, guidelines, and constraints that defines a cohesive and coherent structure consisting of constituent parts, relationships, and connections that establish how those parts fit and work together."*

BODY-5   *"Ultimately, this architectural approach is established to communicate and facilitate the expansion of humans into the universe according to the principles and tenets of NASA's Moon-to-Mars Strategy and Objectives."*

BODY-9   *"Establishing a common architectural language, framework, and integration process to communicate and document the Moon-to-Mars system-of-systems is necessary, and this document is the first step in that process."*

BODY-11   *"Two complementary principles have been developed in the M2M Strategy to address the complex framework: architect from the right and execute from the left. Architecting from the right is described by beginning with the long-term goal (farthest to the right on a timeline) and working backwards from that goal to establish the complete set of elements that will be required for success. Derived from the decomposed plan, systems and elements execute from the left in a regular development process, integrating as systems move left to right within the architecture."*

BODY-13   *"The first step in this process is to define the 'Characteristics and Needs' required to satisfy an objective or a group of objectives. While the objectives themselves focus on desired outcomes, the Characteristics and Needs translate those outcomes into the features or products of the exploration architecture necessary to produce those outcomes. Characteristics and Needs are defined in a form that is still neutral regarding architectural implementation, not specifying a particular solution to produce the desired results, but rather focusing on what is produced or accomplished by the architecture."*

BODY-14   *"In the last step in the decomposition process, the defined Use Cases and Functions are organized to group similar features into representative Reference Missions, Concept of Operations, and Reference Elements. Architecture teams, through trade studies and assessments, develop Reference Elements that can most effectively provide a subset, or group, of the desired Functions within defined constraints. Similarly, teams develop Reference Missions and Concepts of Operations that employ those elements to fulfill the defined Use Cases. This step in the process is the first phase in the development of architectural solutions and demonstrates the viability of the Reference Elements, Reference Missions, and Concept of Operations in delivering the defined Functions and Use Cases, providing the desired Characteristics and Needs, and satisfying the Blueprint Objectives."*

BODY-15   *"In the Architecture Framework, the Sub-Architectures and Segments will be used to ensure coherency in the elements, which may include various programs, projects, or systems, as represented by the lettered and numbered boxes. These programs and projects will be expanded or added to over time with additional elements with which they will need to interface within a sub-architecture. Segments will describe the relationship and cooperation across these elements. As systems mature,* functions may be added or reassigned (denoted as a + or -) to reflect capabilities or implementations through the design or evolution of systems."

BODY-16   "The Communication, Positioning, Navigation, and Timing (CPNT) sub-architecture is a group of services that enable the sending or receiving of information, ability to accurately and precisely determine location and orientation, capability to determine current and desired position, and ability to acquire and maintain accurate and precise time from a standard. Some key factors affecting the implementation of CPNT are the regions in which service is available, the delivery mechanisms for those services to those areas, and the evolution of each aspect throughout the lifetime of the architecture. Another key consideration for a strong foundation is maximizing the interoperability of CPNT assets throughout an evolving architecture with many different providers and users (e.g., government, commercial, scientific, international, etc.)."

BODY-17   "The humans who embark on the exploration missions are the most critical component of the campaign to get humans to the Moon and, ultimately, to Mars. Vehicles, systems, training, and operations must be designed around the 'human system'. In order to provide human-rated systems, standards for design and construction, safety and mission assurance, crew health and performance, flight operations, and system inter-operability are applied."

BODY-18   "The need to deliver elements, payloads, cargo, experiments, and larger quantities of logistics and to better address inventory management, trash, and waste disposal functions necessary to support the missions and meet planetary protection requirements will increase."

BODY-19   "In this document, the term 'utilization' is used generically to encompass all areas of utilization; specific terms, such as 'science or technology demonstration,' are used where the meaning is more specific. The utilization systems sub-architecture is a group of capabilities whose primary function is to accomplish these science, technology, and similar activities, including sample and utilization cargo return to Earth. In this sense, the M2M architecture provides a platform of functions to a broad set of organizations in support of their needs."

BODY-20   "The purpose of segments is to capture at a phase in time the interaction, relationships, and connections of the sub-architectures. These would most commonly be typified by reference missions or operations use cases of the systems to illustrate how systems will work together to achieve objective satisfaction. These examples provide the context for the allocation of functions to elements and systems in the sub-architectures rather than prescriptive solutions. These segments will grow increasingly complex as systems are developed and added to the sub-architectures."

BODY-22   "The SAC trade studies will continue to evaluate concepts and analysis to identify possible solutions to address unallocated functions and potential alternatives. Coordination with both internal NASA and external partner communities will be a key enabler to identify solutions that can most effectively address objective satisfaction. Inputs of technological advancements, alternate concepts, and other innovations can be assessed for satisfaction to meet the integrated architecture needs during the Strategic Analysis Cycles."

BODY-23   "To reach consensus and move forward, an exploration architecture must address all six questions, but reiteration and negotiation may be required. The answer to any one question is less important than ensuring that the answers to all six fit together as an integrated whole."

BODY-24   "Creating a blueprint for sustained human presence and exploration throughout
           the solar system provides a value proposition for humanity that is rooted across
           three balanced pillars: science, inspiration, and national posture."

BODY-25   "The pursuit of scientific knowledge – exploring and understanding the universe – is
           integral to the human space exploration endeavor. Just as the James Webb Space
           Telescope informs about the history of time, answers gained on the Moon and Mars
           will build knowledge about the formation and evolution of the solar system and,
           more specifically, the Earth."

BODY-26   "By implementing an architecture that can be responsive to innovation and
           developments and inclusive of partners, the endeavor will enable benefits reflected
           in terms of both the economy and the human condition."

BODY-27   "These multi-dimensional objectives across the science, technology, and
           infrastructure development goals will need to be supported by foundational
           platforms from which the crew will operate. These systems will provide the crew
           support to retrieve and return samples, deploy instrumentation or technology
           demonstration, research in situ resource utilization, understand the human
           condition in long-term deep space exploration, and much more."

BODY-28   "With an eye to the engineering for systems demand, the lunar South Pole has
           several key driving characteristics to enable systems development for the Moon-to-
           Mars Architecture. First, from a flight performance perspective, the lunar South
           Pole provides a bounding condition for vehicle translation or delta-velocity costs.
           These performance drivers are one of, if not the, most significant condition in
           transportation system design. Vehicles and reference missions designed to achieve
           landing at the South Pole can provide future flexibility to reach global locations
           through planning and certification."

BODY-29   "Having established the NRHO architectural orbit, the ability to transport crew,
           cargo, and support systems to and from the destinations can be decomposed. These
           systems are driven by the sizing performance splits across the architectural
           destinations to traverse the regions from Earth to cislunar space to the surface.
           Crewed transportation systems will be driven by the need to launch, transport, and
           safety mitigate potential contingencies and risks in two key transportation regimes:
           first, crew accessibility to and from Earth to NRHO platforms, and second, to and
           from NRHO to the surface destinations to support either South Pole or non-polar
           mission selection."

BODY-30   "The planned campaign spans a multi-decade period, establishing permanent
           footholds in cislunar space and on the lunar surface, developing and deploying
           major human-rated transportation systems to the Moon and Mars, and developing
           and deploying lunar and Martian surface infrastructure to enable humans to live
           and explore once they arrive. The term 'sustainable' can have different meanings,
           depending on the context. For the exploration campaign, several definitions apply.
           Financial sustainability is the ability to execute a program of work within spending
           levels that are realistic, managed effectively, and likely to be available. Technical
           sustainability requires that operations be conducted repeatedly at acceptable levels
           of risk. Proper management of the inherent risks of deep space exploration is the key
           to making those risks 'acceptable.' Finally, policy sustainability means

BODY-31   "The humans who embark on the exploration missions are the most critical
           component of the campaign to get humans to the Moon, and ultimately to Mars.
           Vehicles, systems, training, and operations must be designed, developed, and
           certified to be safe and reliable for, compatible with, and in support of the 'human

system' as an integrated system to accomplish the mission with an acceptable level of human risk."

BODY-32 "In the five decades since Dr. Wernher von Braun proposed NASA's first human Mars architecture, NASA has pivoted from one exploration point design concept to another, many optimized around heritage programs or emerging technologies of particular interest. Indeed, half a century of architecture studies have filled our libraries with myriad architecture concepts, all having one thing in common: none of these concepts found traction with stakeholders, many of whom had competing perspectives or needs."

BODY-33 "The Apollo program was famously characterized by the mandate of 'landing a man on the Moon and returning him safely to Earth before the end of the decade'. This prioritized 'When?' (within the decade) over other considerations. NASA successfully achieved the goal, but because the resulting architecture was optimized to meet a tight implementation schedule, it was not a particularly extensible architecture, with implications to sustained human exploration of the Moon. The Apollo program serves as a cautionary tale for Mars: if decision-makers focus on 'When?' as an anchoring decision (Figure 2-4), and the answer is a date that does not give us enough time to develop new technologies, then the answer to 'How?' would default to heritage or heritage-derived systems."

BODY-34 "Starting with 'Why?' will help anchor the development process, but architecture choices may still vary widely depending on how the many different answers to 'Why?' are prioritized. Must the first human Mars mission check off every item in the 'Why?' Venn diagram, or is it sufficient to establish a first-mission architecture that meets the highest-priority items, and is extensible to meet lower priorities during subsequent missions?"

BODY-35 "In practice, the Mars architecture decision flow is likely to be iterative rather than linear. To minimize disruption, rework, and cost or schedule changes, understanding the minimum goals and priorities for the first mission, as well as the longer-term goals for subsequent missions, can aid in establishing a flexible and sustainable architecture. The answer to any one of these questions is less important than whether the answers to all six complement one another as a set and can be balanced to establish an architecture that is achievable, affordable, and adaptable."

BODY-36 "To achieve shorter duration roundtrip missions to Mars, less-energy-efficient trajectories must be utilized. The energy vs. time trade for a roundtrip mission to Mars is a continuum, but the relationship is exponential in nature: as the mission duration is shortened, the energy required to achieve the roundtrip mission increases exponentially. This translates to an exponential increase in the vehicle mass required, in terms of both propellant and propulsion system, to achieve the roundtrip journey."

BODY-37 "The farther that humans travel from Earth, the more risk we must accept to achieve the goals of exploration. Mission durations, travel distances, and mass constraints increase the probabilities of something not performing as expected and decrease our ability to respond in a timely manner to emergencies. Crew health, safety, and survival techniques will necessarily change as we move into Mars exploration. The definition of and acceptance of reasonable levels of risk will be a driving factor in determining architecture capabilities and use cases."

BODY-38 "From a purely medical point of view, it would seem intuitively obvious that the 2-year opposition-class mission should be 'better' for the crew than the longer-duration conjunction class mission due to the shorter time spent in the deep space *environment, but that conclusion is premature without more insight into the integrated vehicle risks that will be layered on top of the medical risks, as well as considerations for crew performance."*

BODY-39   *"Address high priority planetary science questions that are best accomplished by on-site human explorers on and around the Moon and Mars, aided by surface and orbiting robotic systems."*

BODY-40   *"Uncover the record of solar system origin and early history, by determining how and when planetary bodies formed and differentiated, characterizing the impact chronology of the inner solar system as recorded on the Moon and Mars, and characterize how impact rates in the inner solar system have changed over time as recorded on the Moon and Mars."*

BODY-41   *"Reveal inner solar system volatile origin and delivery processes by determining the age, origin, distribution, abundance, composition, transport, and sequestration of lunar and Martian volatiles."*

BODY-42   *"Advance understanding of physical systems and fundamental physics by utilizing the unique environments of the Moon, Mars, and deep space." (PPS-2)*

BODY-43   *"Develop the capability to retrieve core samples of frozen volatiles from permanently shadowed regions on the Moon and volatile-bearing sites on Mars and to deliver them in pristine states to modern curation facilities on Earth."*

BODY-44   *"Preserve and protect representative features of special interest, including lunar permanently shadowed regions and the radio quiet far side as well as Martian recurring slope lineae, to enable future high-priority science investigations."*

BODY-45   *"Advance understanding of how physical systems and fundamental physical phenomena are affected by partial gravity, microgravity, and general environment of the Moon, Mars, and deep space transit."*

BODY-46   *"Goal: Create an interoperable global lunar utilization infrastructure where U.S. industry and international partners can maintain continuous robotic and human presence on the lunar surface for a robust lunar economy without NASA as the sole user, while accomplishing science objectives and testing for Mars."*

BODY-47   *Develop and demonstrate an integrated system of systems to conduct a campaign of human exploration missions to the Moon and Mars, while living and working on the lunar and Martian surface, with safe return to Earth.*

BODY-48   *"Goal: Conduct human missions on the surface and around the Moon followed by missions to Mars. Using a gradual build-up approach, these missions will demonstrate technologies and operations to live and work on a planetary surface other than Earth, with a safe return to Earth at the completion of the missions."*

BODY-51   *"The architecture accommodates this approach in the context of available capabilities and differences in the lunar and Mars environments. Initially this is done at the element level, then through combined operations that eventually culminate in several precursor missions in the lunar vicinity where the crew experiences long durations in the deep space environment coupled with rapid acclimation to partial gravity excursions using Mars-like systems and operations. The Mars-forward exploration systems also have the goal to maximize crew efficiency for utilization, which will be tested by a continuum of excursions to a diverse set of sites driven by science needs. The balance between diverse site access and long-duration infrastructure objectives will inform the allocation of functions across systems."*

BODY-52   "The third segment, Sustained Lunar Evolution, is the broad and undefined end state that builds on the foundation of the first two segments and enables capabilities, systems, and operations to support regional and global utilization (science, etc.), economic opportunity, and a steady cadence of human presence on and around the Moon. Here we can envision various uses of the lunar surface and cislunar space to enable science, commerce, and further deep space exploration initiatives."

BODY-53   "The Human Lunar Return (HLR) segment of the exploration campaign includes the inaugural Artemis missions to enable returning humans to the Moon and demonstrating both crewed and uncrewed lunar systems, including the support to initial utilization (science, etc.) capabilities. This segment will be used to demonstrate initial systems to validate system performance and to establish a core capability for follow-on campaign segments."

BODY-56   UC-47 Allow ground personnel and science team to directly engage with astronauts on the surface and in lunar orbit, augmenting the crew's effectiveness at conducting science activities.

BODY-60   "As the first crewed mission returning to the lunar surface, this RM encompasses many use cases that will be repeated throughout the Moon-to-Mars campaign. Starting with transportation, use cases include transporting crew and systems from Earth to cislunar space, staging crewed lunar surface missions from cislunar space, assembling integrated assets in cislunar space, transporting crew and systems between cislunar space and the lunar surface, and returning crew and systems from cislunar space to Earth. The surface portion includes use cases such as crew operations on the lunar surface, frequent crew EVAs on the surface, and crew conducted utilization activities (including science, crew health and performance, and other operations) on the surface and in space."

BODY-61   "The growth of CPNT services throughout the HLR segment will enable the near-term exploration objectives of the HLR segment while providing a robust foundation upon which a scalable infrastructure can grow to support the needs of a sustained lunar presence, including precursor missions that will inform and validate a Martian architecture."

BODY-63   "Logistics represents all equipment and supplies that are needed to support mission activities that are not installed as part of the vehicle. Logistics typically includes consumables (e.g., food, water, oxygen), maintenance items (planned replacement items), spares (for unexpected/unplanned failures), utilization (e.g., science and technology demonstrations), and outfitting (additional systems/sub-systems for the elements), as well as the associated packaging."

BODY-64   "The xEVA System allows crew members to perform extravehicular exploration, research, construction, servicing, repair operations, and utilization and science in cislunar orbit and on the lunar surface. EVA transverse and tasks may be augmented by robotics and rovers. The xEVA System includes the EVA suit, EVA tools, and vehicle interface equipment."

BODY-65   "The SLS is a super-heavy-lift launch vehicle that provides the foundation for human exploration beyond Earth orbit (BEO). With its unprecedented power and capabilities, SLS is the only launch vehicle that can send Orion, astronauts, and payloads directly to the Moon on a single launch. The SLS is designed to be evolvable, which makes it possible to conduct more types of missions, including human missions to Mars; assembly of large structures; and robotic, scientific, and exploration missions to destinations such as the Moon, Mars, Saturn, and Jupiter. Humans will be transported safely, and different payloads will be delivered *efficiently and effectively, to enable a variety of complex missions in cislunar and deep space."*

BODY-66   *"The Orion spacecraft, NASA's next-generation spaceship to take astronauts on a journey of exploration to the Moon and on to Mars, is shown in Figure 3-5."*

BODY-67   *"The LAS, positioned on a tower atop the CM, can activate within milliseconds to propel the vehicle to safety and position the CM for a safe landing."*

BODY-68   *"The initial human landing mission will be a demonstration of this initial HLS configuration and of the minimum basic technologies and innovation required to safely transport crew and utilization cargo to and from the lunar surface."*

BODY-69   *"Investigations and demonstrations launched on commercial Moon flights will help the Agency study Earth's nearest neighbor under the Artemis approach."*

BODY-72   *"Increased mission durations, expanded capabilities, and the ability to access additional regions of the lunar surface enable a growth in utilization, during both crewed and uncrewed mission phases. A variety of science objectives may be addressed during the FE segment, ranging from lunar and planetary science to human and biological science and including science-enabling and applied science goals. During the FE campaign segment, enhanced architecture capabilities would further enhance ability to address and achieve science objectives..."*

BODY-73   *"Objective TH-3 (develop system(s) to allow crew to explore, operate, and live on the lunar surface and in lunar orbit with scalability to continuous presence; conducting scientific and industrial utilization as well as Mars analog activities) drives several characteristics and needs. These include demonstration of capabilities to allow crew to live, to conduct science and utilization activities, and to exit habitable space and conduct EVA activities all in both cislunar space and on the lunar surface."*

BODY-74   *"The Foundational Exploration segment will build on the types of lunar surface exploration accomplished in the HLR segment, which includes crew habitation in the crew lander vehicle with capability for EVA from the lander. In FE, additional use cases may be implemented with the addition of an unpressurized mobility platform to extend EVA range and scientific exploration. This enables the use case for crew excursions to locations distributed around the landing site and has the potential to enable others such as robotic assistance of crew exploration, the locating of samples and resources, and retrieval of samples; crewed/robotic collection of samples from PSRs; and deployment of power generation, storage, and distribution systems at multiple locations around the lunar South Pole, among others."*

BODY-75   *"Working toward the objectives on expanding exploration for longer durations while conducting scientific and industrial utilization, developing surface habitation systems, and performing Mars risk reduction activities prompts the inclusion of additional functional capabilities. With initial surface crew sizes, one method to accomplish these objectives is by adding functionality for pressurized mobility systems. This function may enable use cases such as crew IVA research, additional robotic assistance of crew exploration beyond the unpressurized mobility function, expanded durations for crew operations on the lunar surface (including additional habitation functions), crew excursions to locations distributed around the landing site, EVA egress/ingress, crew/robotic collection of samples from PSRs, and crew relocation and exploration in a shirtsleeve environment."*

BODY-76   *"In order to move beyond the surface mission types in HLR, new capabilities in mobility are necessary to enable exploration beyond the EVA walking range of the* crew. An unpressurized mobility platform or platforms (such as the notional concepts shown in Figure 3-8) may contribute to meeting these needs, whether crewed or uncrewed. Functions that could be grouped into this capability category include providing local unpressurized crew surface mobility, as well as autonomous and/or tele- operations, and enabling additional science and utilization."

BODY-77 *"Expanding surface mission durations and exploration range necessitates new capabilities in crew mobility and habitation. Combining both needs into a pressurized mobility platform or platforms is one option (as seen with the notional concepts in Figure 3-9), potentially encompassing functions such as crew habitation on the lunar surface, crew surface EVAs, pressurized crew surface mobility, crew IVA workspace on the lunar surface, logistics transfer (including fluids and gasses) on the lunar surface, autonomous and/or tele-operations, and mobile crew habitation. Additional functions may include enabling additional science and utilization, such as surface sample (including frozen samples) recovery, curation, and packaging. These surface mobility functions can be accomplished in many ways*

BODY-78 *"Foundational Exploration emphasizes extended duration and preparing for crewed Mars mission profiles through analog missions in lunar vicinity."*

BODY-80 *"In the Sustained Lunar Evolution (SLE) campaign segment, NASA aims to build, together with its partners, a future of economic opportunity, expanded utilization, including science, and greater participation on and around the Moon. The focus on the segment is the growth beyond the Foundational Exploration segment to accommodate objectives of increased global science capability, long duration/increased population, and the large-scale production of goods and services derived from lunar resources. This segment is an 'open canvas,' embracing new ideas, systems, and partners to grow to a true sustained lunar presence."*

BODY-81 *"A sustained architecture at the lunar surface would further enable achievement of key science objectives in Lunar/Planetary Science, Heliophysics, Human and Biological Science, and Physics and Physical Science, as well as facilitate addressing new science objectives identified as a result of discoveries made during the previous campaign segments."*

BODY-82 *"Economic opportunity on and around the Moon in the context of this discussion means that governments are no longer the sole source of support for the funding of the lunar activities and that non-governmental entities would like to invest in, and profit from, activities at the Moon. NASA aims to reduce the barriers of entry for activities on and around the Moon and to provide capabilities others can leverage. Currently there is limited economic rationale for exploring the Moon, but given the cost of getting to and from the Moon, knowledge and access are perhaps the first areas where economic opportunity exists for the non-governmental sector."*

BODY-83 *"Increased science capability, economic opportunity, and duration/population at the lunar South Pole region have the potential to evolve and merge in the future to form the first sustained human civilization beyond Earth."*

BODY-85 *"In truth, mission duration may be thought of as a continuum: the architecture can be optimized for any given duration for a particular opportunity year or a range of durations over different opportunities."*

BODY-86 *"To provide stakeholders with a sense for how the Mars architecture changes as just a single constraint is varied, three reference missions of different total durations— but all with the same surface and transit operational constraints, such as environmental exposure, communication delays and blackout periods —are defined to enable assessment of the architecture to inform the eventual decision* roadmap (Figure 3-14): Reference Mission 0 with an Earth-Mars-Earth transit duration not to exceed 760 days, Reference Mission 1 with a moderate transit duration of 850 days, and Reference Mission 2 with a more relaxed transit duration of up to 1,100 days."

BODY-87   "Is nuclear propulsion needed to enable crewed Mars missions?"

BODY-88   "To aid in assessing extensibility of Mars elements to other destinations or programs and vice versa, the Mars architecture elements can be bucketed into four major categories: 1) Mars surface systems that enable crew to live and work on the planetary surface; 2) Entry, Descent, Landing, and Ascent (EDLA) systems that are able to move crew and surface systems from Mars orbit to the Mars surface, and return crew and cargo back to Mars orbit; 3) transportation systems that are able to move crew and cargo from Earth to Mars orbit and back again; and 4) crew support systems that cross multiple missions, phases, and destinations, such as EVA spacesuits, distributed communications networks, or crew healthcare systems."

BODY-89   "Note that decisions involving the surface mission purpose will have impacts beyond the surface architecture. For example, surface mission purpose will inform landing site selection, which will drive the EDLA, and transportation architectures. Also note that science, technology demonstration objectives, and utilization strategy is, by definition, a key factor in the surface mission purpose decision, with flow-down impacts to landed utilization mass, volume, and power; this factor potentially influences the payload capacity and/or number of landers required."

BODY-90   "Surface mobility decisions will be derived from the mission purpose (where do we need to go to meet the objectives and what do we need to do there?), stay duration (how long do we have to get there and back to the MAV?), cargo movement (what payload elements need to be moved from one location [e.g., the lander deck] to other locations?), and habitation decisions (are the habitable volumes moveable? what are the traverse distances to/from habitat, landing site, and ascent stage?). Each of these individual considerations will influence the overall exploration radius."

BODY-91   "Advanced generation of Ingenuity, a robotic helicopter landed with Perseverance and currently in use on Mars."

BODY-93   "Due to a combination of potential crew deconditioning, lengthy communications delays with Earth, and the rapid pace of dynamic events during EDL, Mars EDL systems must be designed for autonomous operation with limited real-time crew input."

BODY-94   "The largest payload landed to date on Mars is about one metric ton, but even the most modest human Mars mission is estimated to require at least 75 t of total landed payload for even a short-duration surface stay. Longer, more ambitious missions will require more landed mass."

BODY-95   "Reuse cannot be an afterthought for EDLA systems. It must be integral to the design. Feasibility of reusing EDLA systems is highly coupled between system design and concept of operation. Initial reference designs are not practical for reuse, but with changes to design and operation, reuse could be enabled. Certain designs may be more 'evolvable' for reusability than others."

BODY-97   "Selection of a human Mars transportation system will be a complex decision shaped by numerous factors, such as mission objectives (the 'Why?' question), exploration partner contributions and commitments, programmatic schedules, and integrated risk assessments. The four transportation architectures presented here represent the range of options currently being analyzed."

BODY-98   *"The total roundtrip mission duration for a Mars mission is the primary driver for any in-space transportation decisions. Longer mission durations (~3 years) typically require lower energy, as they can rely on the more favorable alignments between Earth and Mars to perform two optimal transfers between the planets. Shorter missions would require more energy to complete, as the in-space transportation system will need to complete the roundtrip mission while fighting against the natural orbital energy of the two planets. The energy required, and therefore the propulsion technology and total propellant mass, scales exponentially with mission duration, so the shorter missions are exponentially harder than the longer missions."*

BODY-99   *"The selection of the parking orbit at Mars for staging and aggregation of the mission will be dependent on the architecture and mission mode decisions, as well as surface abort timing constraints. Current assumption for Mars parking orbit is a 5-sol orbit, with the perigee of the parking orbit directly above the landing site to support a direct landing. This high-altitude parking orbit is beneficial to the transportation system because it does not require the whole transportation stack to insert deep into Mars' gravity well, but it puts an additional burden on the MAV, as the energy and time required to reach 5-sol orbit is higher than for a lower parking orbit."*

BODY-100   *"The reusability of any of the transportation elements is a key driver in the design of the system. If additional follow-on missions to Mars are desired to establish routine access to Mars' surface, then the ability to reuse elements will be a key decision in enabling these missions."*

BODY-101   *"To better understand the performance of various propulsion system designs in the context of the analysis reference missions, four different propulsion and power options are currently under evaluation: a hybrid Nuclear Electric Propulsion (NEP)/Chemical Propulsion system, Nuclear Thermal Propulsion (NTP) System, hybrid Solar Electric Propulsion (SEP)/Chemical Propulsion System, and All-Chemical Propulsion Systems."*

BODY-103   *"As shown in Figure 3-16, Figure 3-17, Figure 3-18, and Figure 3-19, several conceptual designs of each transportation architecture are being developed to allow stakeholders to better assess option performance across the range of mission duration options. These figures demonstrate how vehicle size and complexity vary as just one parameter, total mission duration, is varied."*

BODY-105   *"A unique challenge for a Mars mission will be addressing the approximately two-week period during which the Sun interrupts the line-of-sight path between the crew and Earth, and no direct communication is possible. An uninterruptable relay could mitigate this blackout period, though it should be noted that this potentially increases the communications lag time, since the relay must be placed far enough from Mars to maintain line of sight to Earth when the Sun is between Earth and Mars."*

BODY-106   *"The Mars communications system concept of operations will be substantially different from lunar operations due to the delay caused by the increased distance from Earth-based ground support, up to 22 minutes each way, and the annual communications blackout of up to two weeks. The current architecture concepts posit the Mars transit vehicle acting as the primary relay between the surface systems network to Earth-based networks during the crewed surface phase of the mission. Another concept under consideration is having the surface crew relying on the orbital crew (that remain aboard the Mars transportation system) to provide low-latency verbal guidance and expertise to augment the longer latency-Earth support. It remains forward work to fully develop the Mars communications concept of operations that supports both near- and far-range operations with varying magnitudes of latency*

BODY-108   *"Current concepts assume the suit can egress and ingress habitable vehicles and provide life support, thermal control, protection from the environment, communications, and mobility/dexterity features designed to interact with spacecraft interfaces and supporting tools and equipment such that exploration, science, construction, and vehicle maintenance tasks can be done safely and effectively."*

BODY-109   *"Disposal of trash is a key issue for the in-space portion of the Mars mission. Because of the propulsive requirements, it is undesirable to accumulate trash in the TH. Methods to dispose of trash during the transits to and from Mars will be considered to reduce the total TH mass."*

BODY-110   *"Crew systems for habitability include direct crew care systems such as food and nutrition consumables and preparation equipment, personal hygiene systems including body waste management, clothing, housekeeping equipment and consumables, physiological countermeasure systems (such as aerobic and resistance exercise equipment), crew privacy systems and accommodations conducive for sleep."*

BODY-111   *"An integral part of the Moon-to-Mars architecture is the desire to usher in a new era of exploration with the recognition of the mutual interest between NASA and international partners in the exploration and use of outer space for peaceful purposes. Coordination and cooperation between and among established and emerging actors in space is a foundational principle of Artemis. This is best accomplished through partnership and collaboration with members of the global community and should be reflected in every segment of the Moon-to-Mars Architecture."*

BODY-112   *"Cooperation will occur across the full spectrum of opportunities–from infrastructure to science, technology, and education activities on and around the Moon, Mars, and beyond. New opportunities for cooperation will emerge as the architecture is further developed each year and collaboration is discussed between NASA and its prospective international partners. International cooperation will advance broad science, exploration, and space technology goals and objectives, as well any number of objectives related to education, inspiration, and public engagement."*

BODY-113   *"It is important to recognize the iterative and ongoing nature of international discussions on architecture and our Moon-to-Mars partnerships. NASA has engaged and will continue engaging with international counterparts, both bilaterally and multilaterally. Along with bilateral discussions, some highlighted above, multilateral discussions have demonstrated utility in advancing mutual understanding and common exploration interests."*

BODY-114   *"The Moon-to-Mars architecture treats the safety of the crew as an utmost concern. However, significant knowledge gaps exist relative to the adverse effects of long-term exposure to the deep space environment. The architecture will be developed to account for known health and medical concerns with deep space missions, as well as for contingency scenarios for failures in mission elements and systems. Significant knowledge has been gained in the on-going human health research aboard the International Space Station and will continue with lunar orbital and surface missions. Long duration Mars precursor missions conducted in cislunar space and on the lunar surface will provide some knowledge and experience. Likewise,* knowledge of the reliability gaps with mission hardware, software, and operations will be tested and refined based on knowledge of LEO missions."

BODY-115   "Maximizing crew time is a critical driver for the exploration architecture across all segments of the campaign. Specifically, this refers to crew time made available for utilization activities, separate from other crew time allocations such as maintenance time. During the HLR campaign segment, the architecture and reference missions emphasize crew exploration by EVA on the lunar surface. This is enabled by allocating functions to the elements in this phase such that maintenance and construction overhead activities are minimized. Concurrently, utilization activities at the Gateway are conducted in cislunar space to complement the surface exploitation activities."

BODY-116   "Additionally, the responsible use of the Moon-to-Mars architecture may require deeper scrutiny of cultural and societal implications of future exploration."

BODY-117   "The success of the Moon-to-Mars architecture is dependent on leveraging the past and current human and robotic spaceflight experience to inform future system design and operational needs. The Moon-to-Mars architecture builds upon and leverages experience gained from past and present spaceflight experience."

BODY-118   "Uncover the record of solar system origin and early history, by determining how and when planetary bodies formed and differentiated, characterizing the impact chronology of the inner solar system as recorded on the Moon and Mars, and characterize how impact rates in the inner solar system have changed over time as recorded on the Moon and Mars."

BODY-119   "Advance understanding of the geologic processes that affect planetary bodies by determining the interior structures, characterizing the magmatic histories, characterizing ancient, modern, and evolution of atmospheres/exospheres, and investigating how active processes modify the surfaces of the Moon and Mars."

BODY-121   Reveal inner solar system volatile origin and delivery processes by determining the age, origin, distribution, abundance, composition, transport, and sequestration of lunar and Martian volatiles.

BODY-124   "Improve understanding of magnetotail and pristine solar wind dynamics in the vicinity of the Moon and around Mars. Understand the effects of short- and long-duration exposure to the environments of the Moon, Mars, and deep space on biological systems and health, using humans, model organisms, systems of human physiology, and plants."

BODY-126   "Conduct astrophysics and fundamental physics investigations of deep space and deep time from the radio quiet environment of the lunar far side."

BODY-127   "Develop the capability to retrieve core samples of frozen volatiles from permanently shadowed regions on the Moon and volatile-bearing sites on Mars and to deliver them in pristine states to modern curation facilities on Earth."

BODY-156   "Human-rating is the process of designing, evaluating, and assuring that the total system can safely conduct the required human missions. Human-rating includes the incorporation of design features and capabilities that accommodate human interaction with the system to enhance overall safety and mission success. Human-rating includes the incorporation of design features and capabilities to enable safe recovery of the crew from hazardous situations."

NASA/TP-20230002706

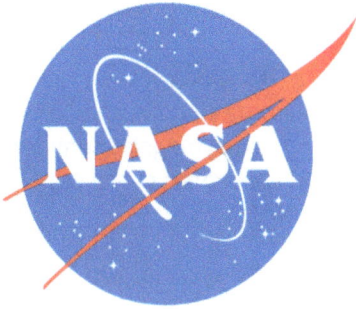

Exploration Systems Development
Mission Directorate

Moon-to-Mars Architecture Definition Document
(ESDMD-001)

NASA STI Program Report Series

The NASA STI Program collects, organizes, provides for archiving, and disseminates NASA's STI. The NASA STI program provides access to the NTRS Registered and its public interface, the NASA Technical Reports Server, thus providing one of the largest collections of aeronautical and space science STI in the world. Results are published in both non-NASA channels and by NASA in the NASA STI Report Series, which includes the following report types:

- TECHNICAL PUBLICATION. Reports of completed research or a major significant phase of research that present the results of NASA Programs and include extensive data or theoretical analysis. Includes compilations of significant scientific and technical data and information deemed to be of continuing reference value. NASA counterpart of peer-reviewed formal professional papers but has less stringent limitations on manuscript length and extent of graphic presentations.

- TECHNICAL MEMORANDUM. Scientific and technical findings that are preliminary or of specialized interest, e.g., quick release reports, working papers, and bibliographies that contain minimal annotation. Does not contain extensive analysis.

- CONTRACTOR REPORT. Scientific and technical findings by NASA-sponsored contractors and grantees.

- CONFERENCE PUBLICATION. Collected papers from scientific and technical conferences, symposia, seminars, or other meetings sponsored or co-sponsored by NASA.

- SPECIAL PUBLICATION. Scientific, technical, or historical information from NASA programs, projects, and missions, often concerned with subjects having substantial public interest.

- TECHNICAL TRANSLATION. English-language translations of foreign scientific and technical material pertinent to NASA's mission.

Specialized services also include organizing and publishing research results, distributing specialized research announcements and feeds, providing information desk and personal search support, and enabling data exchange services.

For more information about the NASA STI program, see the following:

- Access the NASA STI program home page at http://www.sti.nasa.gov

- Help desk contact information:

https://www.sti.nasa.gov/sti-contact-form/

and select the "General" help request type.

NASA/TP-20230002706

Exploration Systems Development Mission Directorate

Moon-to-Mars Architecture Definition Document (ESDMD-001)

National Aeronautics and
Space Administration

Mary W. Jackson Headquarters
Washington, D.C.

April 2023

EXECUTIVE SUMMARY

The National Aeronautics and Space Administration's mission is to explore the unknown in air and space, innovate for the benefit of humanity, and inspire the world through discovery. Key in this mission is extending the reach of humanity through the human exploration of the Moon, Mars, and beyond. To enable this effort, long-term goals and objectives have been established in the Moon-to-Mars Strategy and Objectives document; however, the practical management and execution to ensure objective satisfaction requires an innovative approach to the definition of NASA's Moon-to-Mars human exploration architecture. Architecture is the high-level unifying structure that defines a system. It provides a set of rules, guidelines, and constraints that defines a cohesive and coherent structure consisting of constituent parts, relationships, and connections that establish how those parts fit and work together. This Architecture Definition Document establishes the process, framework, and decomposition of objectives to empower the executing systems', programs', and projects' success in achieving human exploration of the cosmos.

As established in the NASA's Moon-to-Mars Strategy and Objectives, "Why" we explore encompasses three pillars: Science, Inspiration, and National Posture. Ensuring success in all three areas requires an architectural approach incorporating innovation, collaboration, and partnerships that can be sustained across a multi-decadal effort. The Moon-to-Mars human exploration architecture approach decomposes the blueprint objectives into the Characteristics and Needs necessary to satisfy the objectives as they apply to the human exploration systems and supporting elements. These are then traced through Use Cases and Functions to assigned programs, projects, and systems to ensure clear context throughout execution and development. Within the Moon-to-Mars Architecture, the elements will be coordinated via this framework of sub-architectures and campaign segments to ensure thorough decomposition. This framework will ensure architecture progression that increases objective satisfaction through campaign segments to return humans to deep space, explore the environment, and sustain their presence in deep space. This constant traceability and iteration through the architecture process between the current state of execution and future goals and desired outcomes will allow the infusion of technology, innovation, and partnerships.

The first Moon-to-Mars campaign segment, Human Lunar Return, establishes the initial capabilities, systems, and operations necessary to re-establish human presence on and around the Moon. It captures the missions that will test NASA's deep-space crew and cargo transportation system (Space Launch System, Orion, Exploration Ground Systems), implement the initial portion of Gateway to support the lunar missions, deploy and establish a lunar orbital communications relay, exercise the Human Landing System by bringing two crew members to the lunar surface for ~6 days each year, and explore the lunar surface in search of science and resource sites of interest. This effort is instrumental in demonstrating key features of the architecture derived from the exploration objectives, such as cislunar aggregation, safe and reliable crew transportation, and integrated operations extended to the lunar environment.

Following the return to the Moon, the Foundational Exploration segment will expand operations, capabilities, and systems to enable complex orbital and surface missions capable of extensive scientific and technological utilization and Mars forward precursor missions. As these systems expand, the enterprise will provide the technologies and growth to continue a Sustained Lunar Evolution segment, thereby enabling operations, capabilities, and systems to support regional and global utilization, economic opportunity, and a steady cadence of human presence on and around the Moon. Throughout this architectural approach, the continual development and incremental progress will be measured, assessed, and matured to facilitate the Humans to Mars segment, including the initial capabilities, systems, and operations necessary to support it and continued exploration beyond Mars. Ultimately, this architectural approach is established to communicate and facilitate the expansion of humans into the universe according to the principles and tenets of NASA's Moon-to-Mars Strategy and Objectives.

REVISION AND HISTORY

The NASA Office of Primary Responsibility for this document is the Exploration Systems Development Mission Directorate Architecture Development Office. Please visit https://www.nasa.gov/MoonToMarsArchitecture for the latest version and updates to the Moon-to-Mars architecture and exploration campaign.

Revision No.	Description	Release Date
Initial	Initial Release (Reference NASA/TP-20230002706)	04/18/2023

TABLE OF CONTENTS

1.0   INTRODUCTION

Architecture is the high-level unifying structure that defines a system. It provides a set of rules, guidelines, and constraints that define a cohesive and coherent structure consisting of constituent parts, relationships, and connections that establish how those parts fit and work together. This definition, as found in the National Aeronautics and Space Administration (NASA) Systems Engineering Handbook,[1] is essential to capture the broad range of systems, programs, and projects supporting the human exploration of the Moon, Mars, and beyond. Although this definition is typically used for a single program construct rather than a multidecadal Moon-to-Mars (M2M) human exploration architecture, the need for a unifying structure to address the magnitude of the endeavor remains. These goals represent the most complex systems engineering effort conducted by NASA to date. Ultimately, the programs, projects, and contributing systems will span decades, agencies, countries, cultures, and a variety of commercial, academic, and other types of contributors. Establishing a common architectural language, framework, and integration process to communicate and document the Moon-to-Mars system-of-systems is necessary, and this document is the first step in that process.

1.1   PURPOSE

There are many opportunities in executing the ambitious Moon-to-Mars efforts through an integrated architecture. NASA addresses this in its Moon-to-Mars Strategy and Objectives Development[2] document (hereafter referred to as the M2M Strategy). Many of these opportunities involve establishing a system engineering framework that can support the breadth of necessary program and system contributions. By applying these needs to nearer-term lunar development, NASA will be instituting the process, procedures, and techniques needed to enable longer-term Mars goals and more. Some of the challenges being addressed in the M2M Strategy are associated with the architecture definition and include broad/changing goals, funding, and external pressures/influences. This document and the methodology outlined for architecture definition have been crafted to contend with these using an iterative and adaptable framework.

The primary purpose of the Architecture Definition Document (ADD) is to capture the methodology, organization, and decomposition necessary to translate the broad objectives outlined in the M2M Strategy into functions and use cases that can be allocated to implementable programs and projects. Inherent in this process will be the need to communicate the long-term vision, maintain traceability to responsible parties, and iterate on the architectural implementation as innovations and solutions develop. This document will be updated and improved in conjunction with the Architecture Concept Review (ACR) which will be held annually to help unify the Agency and to get buy-in and input from across the Agency on the human exploration architecture. The annual nature of the process provides the opportunity to continually incorporate new developments in technologies and new partnerships, whether they be with industry, the U.S. Government, international entities, or academia.

1.2   SCOPE

The scope of this document is to capture the programs, projects, systems, and contributions that enable the human exploration of the Moon, Mars, and beyond. The Agency level M2M Strategy encompasses the combined objectives that may be satisfied through human, robotic, or other

---

[1] NASA System Engineering Handbook. SP-20170001761.
[2] NASA's Moon-to-Mars Strategy and Objectives Development. NP-2023-03-3115-HQ.

efforts conducted by all Agency directorates. This ADD, the methodology, and the decomposition of the objectives has been conducted by ESDMD for those applicable to the human exploration architecture and robotic systems interfacing or supporting it. Agency Blueprint Goals and Objectives will, in many cases, also decompose or be supported by independent robotic or other non-NASA systems that, in combination with the human architecture, contribute to complete objective satisfaction. Objective decompositions in the ADD identify those derived to support human exploration architecture and systems and may also have other functions, features, or uses beyond those presented here. The Moon-to-Mars Architecture process will coordinate objective decomposition in conjunction with all NASA Mission Directorates.

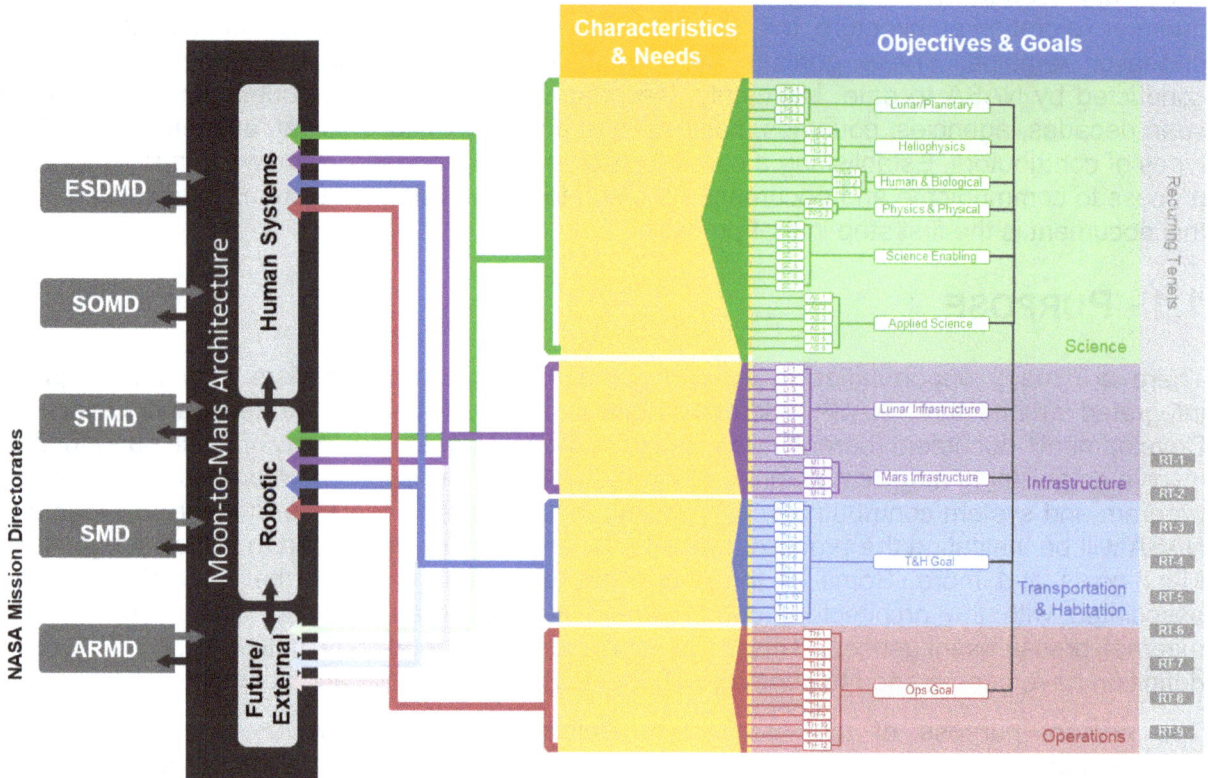

Figure 1-1. Human Exploration Moon-to-Mars Architecture Scope

This ADD has been laid out to reflect the process and will be iterated on over time through subsequent analysis and integration efforts with partners. This Section 1.0 includes description of the methodology and framework of the decomposition. This description includes definitions of the Segments and Sub-Architectures used to describe the architecture and the process by which NASA will organize the decomposition through iterative cycles.

Section 2.0 Architecture Decomposition includes the rationale for the lunar architecture as viewed through a system engineering lens. This describes the key drivers and questions that must be answered to arrive at the implemented architecture. Unique considerations for the Moon are also included. This section introduces the relationships of the questions and how the order they are answered in drives the Mars architecture. This content will eventually be replaced by the Mars architecture description as decisions are made and implemented. This section concludes with the decomposition of the objectives to the characteristics and needs the architecture must possess to support the M2M Strategy and Objectives.

Section 3.0 Moon-to-Mars Architecture includes the relationship of the Characteristics and Needs to the assigned Use Cases and Functions as applied to supporting architecture elements. These elements are organized by the architecture framework introduced in Section 1.0. This section also identifies open or unanswered questions in the architecture and the unallocated functions that are yet to be addressed by future systems or supporting elements. Descriptions of open trades or considerations for future architecture development are included in particular for the Mars architecture.

Section 4.0 Assessment to the Recurring Tenets provides assessments of the architecture and reflects on the degree to which the architecture is adhering to the cross-cutting tenets of the Strategy and Objectives. These assessments are qualitative in nature to consider the state of the architecture and identify opportunities for revision. These will be living assessments updated on a recurring basis as the architecture adapts and develops.

The document content is followed by extensive decomposition and traceability tables in APPENDIX A: Full Decomposition of Lunar Objectives. This Appendix provides the complete traces from lunar Objectives to the implementing element lunar Use Case and Functions. Appendix B provides Acronyms and Abbreviations and Glossary of Terms for reference.

1.3 ARCHITECTURE METHODOLOGY

Two complementary principles have been developed in the M2M Strategy to address the complex framework: architect from the right and execute from the left. Architecting from the right is described by beginning with the long-term goal (farthest to the right on a timeline) and working backwards from that goal to establish the complete set of elements that will be required for success. Derived from the decomposed plan, systems and elements execute from the left in a regular development process, integrating as systems move left to right within the architecture.

An applied systems engineering method has been developed to facilitate applying these principles to the architecture definition. The first part of this method is an ordered process of objectives' decomposition to complete the process of architecting from the right. In this process, the characteristics and needs are identified to assure objective satisfaction. These characteristics and needs are then traced to the functions and use cases that must be accomplished by elements and systems. The second supporting method is establishing an architectural framework to organize, integrate, and track the allocation of functions and use cases to the executing programs and projects. This structure will enable the integration of the system-of-systems development, identify gaps in the architecture, and adjust the architecture as left-to-right execution occurs, technologies mature, or objectives are satisfied. The architectural framework will be managed using sub-architectures and segments, which will be discussed in Sections 1.3.2.1 and 1.3.2.2 respectively.

1.3.1 Objective Decomposition Process

As documented in the M2M Strategy, the broad top-level objectives of the Moon-to-Mars campaign have been identified with the help of stakeholders. These objectives establish desired results for NASA's exploration activities, with each objective defining a desired outcome of the Moon-to-Mars Architecture. Objectives were purposely drafted to be agnostic with respect to implementation. Objectives do not specify architectural or operational solutions. Rather, they provide the goals to facilitate the development of an architecture and the means to measure progress.

To facilitate the objective decomposition process, several terms are defined as follows:

Table 1-1. Key Architecture Process Terms and Definitions

Architecture	The high-level unifying structure that defines a system. It provides a set of rules, guidelines, and constraints that define a cohesive and coherent structure consisting of constituent parts, relationships, and connections that establish how those parts fit and work together.[3]
Needs	A statement that drives architecture capability, is necessary to satisfy the Moon-to-Mars Objectives, and identifies a problem to be solved, but is not the solution.
Characteristics	Features or activities of exploration mission implementation necessary to satisfy the Goals and Objectives.
Use Cases	Operations that would be executed to produce the desired Needs and/or Characteristics.
Functions	Actions that an architecture would perform to complete the desired Use Case.
Segments	A portion of the architecture, identified by one or more notional missions or integrated Use Cases, illustrating the interaction, relationships, and connections of the sub-architectures through progressively increasing operational complexity and objective satisfaction.
Sub-Architecture	A group of tightly-coupled elements, functions, and capabilities that perform together to accomplish architecture objectives.

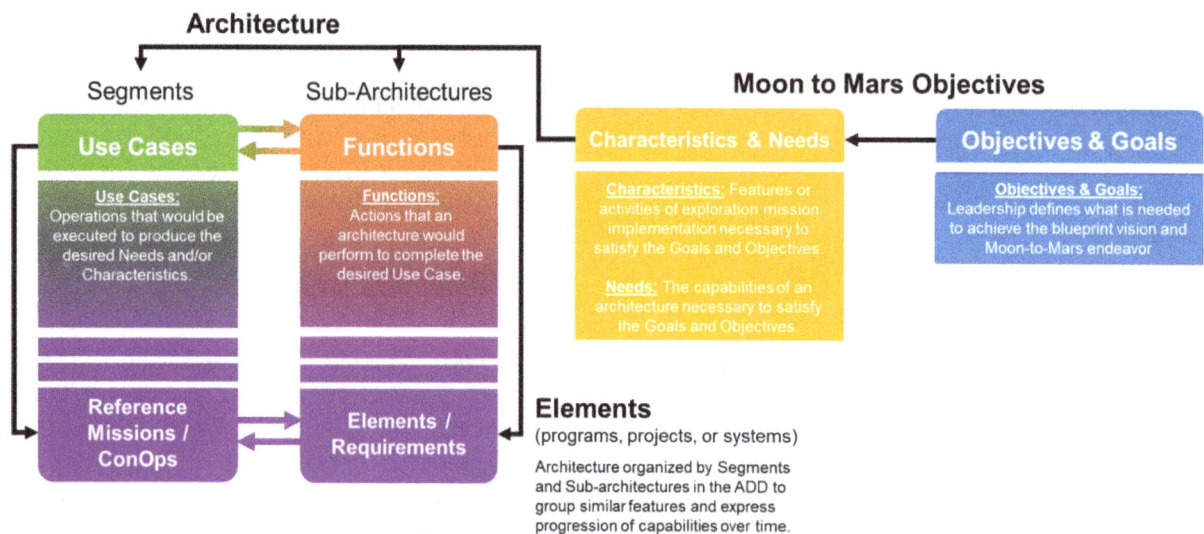

Figure 1-2. Objective Decomposition Process

The process that NASA will apply to define the exploration architecture, described in Figure 1-2, is rooted in the defined set of Top-Level Objectives within the M2M Strategy. The process includes

---

[3] Definition from NASA System Engineering Handbook. SP-20170001761.

a series of discrete steps, each of which results in the progressive definition of needs with reduced abstraction in the architecture and an increasing level of fidelity.

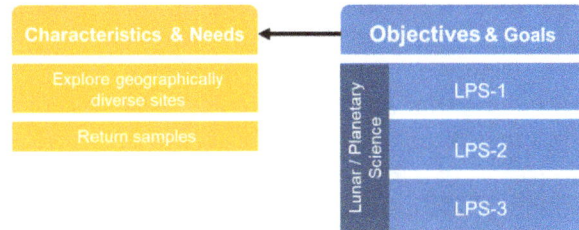

Figure 1-3. Notional Example Mapping of an Objective to Characteristics and Needs

The first step in this process is to define the "Characteristics and Needs" required to satisfy an objective or a group of objectives. While the objectives themselves focus on desired outcomes, the Characteristics and Needs translate those outcomes into the features or products of the exploration architecture necessary to produce those outcomes. Characteristics and Needs are defined in a form that is still neutral regarding architectural implementation, not specifying a particular solution to produce the desired results, but rather focusing on what is produced or accomplished by the architecture. This step of the process is critical in taking generalized objectives and converting them into actionable exploration activities. Goal owners and stakeholders who are familiar with and helped to define the Top-Level Objectives within the M2M Strategy contribute to the definition of the Characteristics and Needs, adding the detail needed to define the features and products. Figure 1-3 shows a partial and notional example of how one representative objective could be decomposed into a set of Characteristics and Needs.

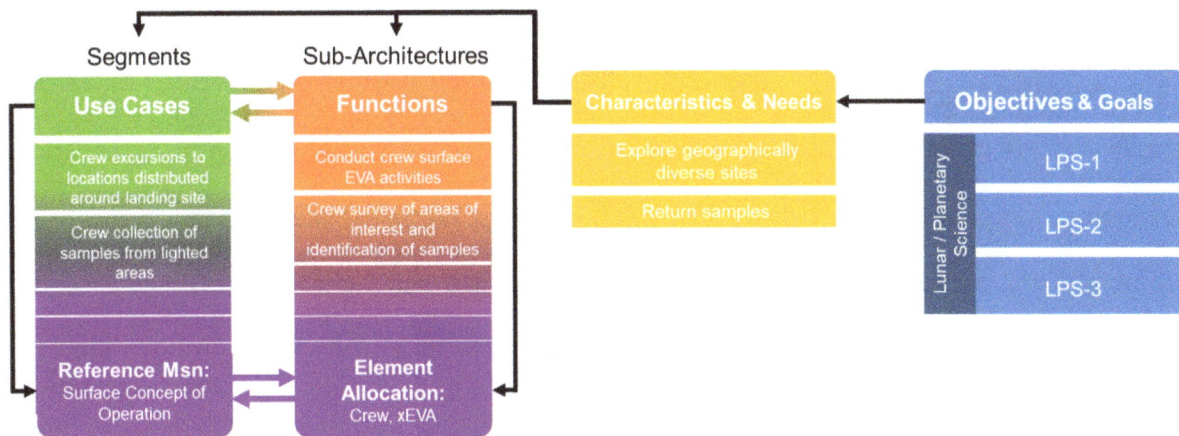

Figure 1-4. Notional Example Mapping of Characteristics and Needs to Functions and Use Cases

Once the Characteristics and Needs are defined, the next step in the process is to translate those statements into a more specific definition of implementable "Functions" and "Use Cases." This step adds further definition to the architectural needs and begins to define actionable features that could be included in the exploration architecture. Functions are the services or actions that would have to be produced by the exploration architecture to provide the desired Characteristics and Needs. Use Cases describe how those functions are operationally employed to produce the desired Characteristics and Needs. Architecture teams formally decompose the Characteristics and Needs into Functions and Use Cases, working with stakeholders to ensure that the defined Functions and Use Cases would result in the desired outcomes.

In the last step in the decomposition process, the defined Use Cases and Functions are organized to group similar features into representative Reference Missions, Concept of Operations, and Reference Elements. Architecture teams, through trade studies and assessments, develop Reference Elements that can most effectively provide a subset, or group, of the desired Functions within defined constraints. Similarly, teams develop Reference Missions and Concepts of Operations that employ those elements to fulfill the defined Use Cases. This step in the process is the first phase in the development of architectural solutions and demonstrates the viability of the Reference Elements, Reference Missions, and Concept of Operations in delivering the defined Functions and Use Cases, providing the desired Characteristics and Needs, and satisfying the Blueprint Objectives. Figure 1-4 shows an example of how the notional Characteristics and Needs could be further decomposed into notional Functions and Use Cases. The decomposition of Blueprint Objectives is provided in Appendix A and will continue to be refined during future process cycles.

From the Use Cases and Functions, the definition of Design Reference Missions, Concepts of Operations, and System Requirements can be traced. Through the program or project formulation process, the allocated Use Cases and Functions will be used to address feasibility, definition, and scope. Programmatic assessments will identify the existence of feasible solutions to meet the assigned Functions and Use Cases as requirements are instantiated. If adjustments are needed in formulation, Functions/Use Cases may be descoped for later allocation in the architecture process to a different system. During design and development, assessments will be conducted to ensure the system is achieving the expected architectural functions or adjustments are made as needed. Groupings and definitions may change as designs progress and/or are better understood; however, it is important that the mapping of objectives to Reference Missions, Concept of Operations, and Systems be continually revisited to assure objective satisfaction as intended.

1.3.2 Architecture Framework

Given the scale of the Moon-to-Mars architecture, it is necessary to establish a framework for partitioning the effort into portions executable by NASA and its partners. Instituting a system engineering process that empowers incremental advancements and the ability to infuse innovations in technologies and solutions provides the opportunity for economic benefit and the incorporation of partnerships while ensuring objectives are systematically accomplished. In a typical system engineering process, the architecture would be fully established up front, requirements and concept of operations defined, and the programs would begin execution. This traditional method, if applied to the scale of Moon-to-Mars architecture, would therefore have to "pick" the mission profile, technologies, and development schedule for an enormous number of projects up front and would bias solutions to mature solutions and capabilities that exist today. This traditional "single pass through" architecture definition has been attempted for Moon and Mars systems many times in the past with limited success, as discussed in the M2M Strategy document.

To contend with this architecture breadth, an iterative framework process is established using two types of integration categories. The first type is to group tightly coupled systems, needs, and capabilities that function together to accomplish objectives as Sub-Architectures, similar to a system-to-sub-system relationship. More detail on the Sub-Architectures can be found in Table 1-2. The second type is to establish Segments defined as a portion of the architecture, identified by one or more notional missions or integrated use cases, illustrating the interaction, relationships, and connections of the Sub-Architectures through progressively increasing operational complexity and objective satisfaction. More detail on the specific Segments will be discussed in Section 3.0 and Table 3-1 specifically. Segments reflect the integration reference missions established to ensure elements can function together. Actual missions and Segments operations may overlap and do not necessitate conclusion of one before functions and projects in the next begin operations. Together, these provide horizontal (Sub-Architecture) and vertical (Segment) integration to provide traceability in the Moon-to-Mars architecture definition as illustrated in Figure 1-5.

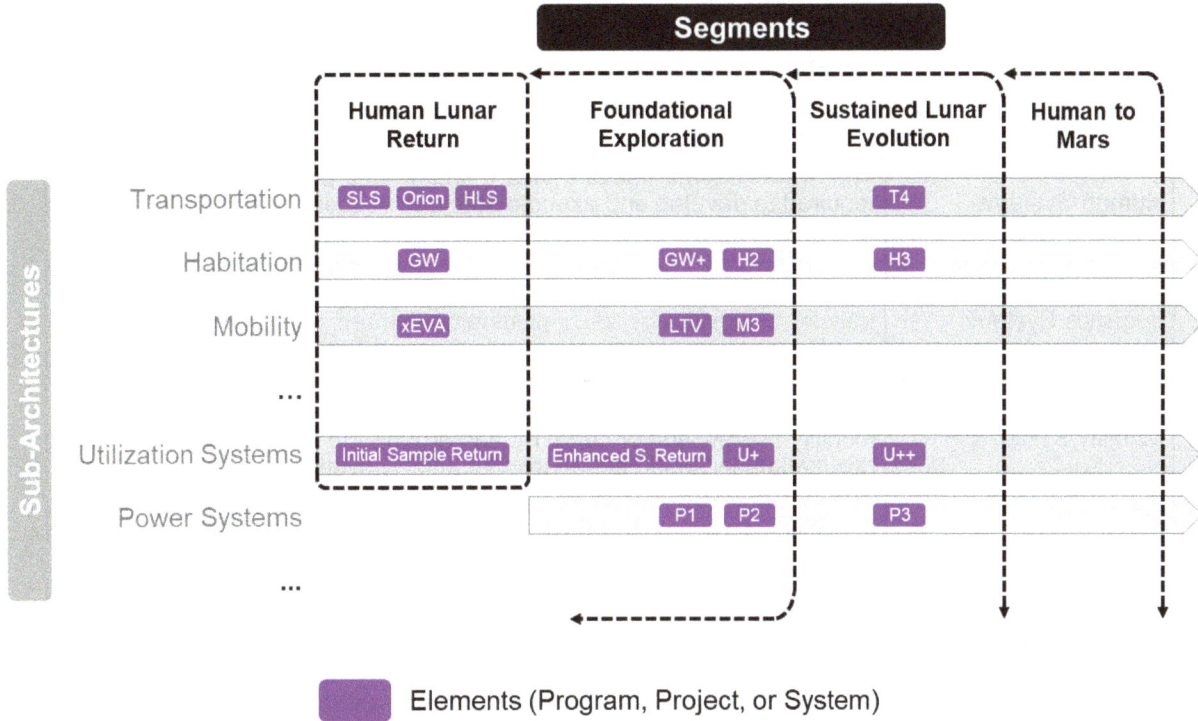

Figure 1-5. Illustration of the Moon-To-Mars Architecture Framework, (+/- denote added or reassigned functions)

In the Architecture Framework, the Sub-Architectures and Segments will be used to ensure coherency in the elements, which may include various programs, projects, or systems, as represented by the lettered and numbered boxes. These programs and projects will be expanded or added to over time with additional elements with which they will need to interface within a sub-architecture. Segments will describe the relationship and cooperation across these elements. As systems mature, functions may be added or reassigned (denoted as a + or -) to reflect capabilities or implementations through the design or evolution of systems.

1.3.2.1   Sub-Architecture Definitions

The use of Sub-Architectures address the complexity of programs, projects, systems, and operations that span multiple sources or elements but must interact in a tightly coupled manner. By sub-dividing the architecture, functions and use cases can be assessed for consistency, gaps, or improvements. These sub-architectures will then evolve through the ADD iterations as functions and use cases are assigned to associated elements and systems to facilitate increasing capabilities toward the accomplishment of objectives. As shown in Figure 1-5, sub-architectures will add elements and systems through the progression of segments to achieve the associated characteristics and needs. These sub-architectures can facilitate and identify the areas where common standards and inter-operability of associated elements is beneficial to ensure consistency in functions and allocations.

Table 1-2. Sub-Architecture Definitions

Communication, Positioning, Navigation, and Timing Systems	A group of services that enable the sending or receiving of information, ability to accurately and precisely determine location and orientation, capability to determine current and desired position, and ability to acquire and maintain accurate and precise time from a standard.
Habitation Systems	A group of capabilities that provide controlled environments to ensure crew health and performance.
Human Systems	The overall capabilities of onboard and ground personnel and systems required to develop and execute safe and successful crewed and uncrewed missions.
Logistics Systems	Systems and capabilities needed for packaging, handling, staging, and transfer of logistics goods, including equipment, materials, supplies, and Environmental Control and Life Support System consumables.
Mobility Systems	A group of capabilities and functions that enable the robotic-assisted mobility of crew and/or cargo on and around the surface of the destination, including extravehicular activity systems.
Power Systems	Capabilities that support the function of providing electrical energy to architectural elements. These capabilities include components and hardware for power generation, power conditioning and distribution, and energy storage.
Transportation Systems	Capabilities that provide the transportation functions for all phases of the Moon and Mars missions for both crew and cargo, including in-space, Entry, Descent, and Landing (EDL), and ascent for all Earth, Moon, and Mars phases.
Utilization Systems	A group of capabilities whose primary function is to accomplish utilization which enables science and technology demonstrations.

The initial set of identified sub-architectures reflects the current state of program and project development and identified integration challenges. While the sub-architectures are defined independently, they will have interfaces and dependencies with other sub-architectures and will all work together to perform utilization activities supported by the architecture. It is expected that refinement of the current sub-architectures and identification of additional ones will occur in the Architecture Concept Review Cycles. The initial sub-architectures are identified, and the rationale is provided in Table 1-2.

1.3.2.1.1 Communication, Positioning, Navigation, and Timing Systems

The Communication, Positioning, Navigation, and Timing (CPNT) sub-architecture is a group of services that enable the sending or receiving of information, ability to accurately and precisely determine location and orientation, capability to determine current and desired position, and ability to acquire and maintain accurate and precise time from a standard. Some key factors affecting the implementation of CPNT are the regions in which service is available, the delivery mechanisms for those services to those areas, and the evolution of each aspect throughout the lifetime of the architecture. Another key consideration for a strong foundation is maximizing the interoperability of CPNT assets throughout an evolving architecture with many different providers and users (e.g., government, commercial, scientific, international, etc.). As the architecture evolves, the CPNT sub-architecture and concept of operations will scale based on the developing user needs and will evolve by collecting ground truth data as the campaign progresses. Services will expand (for example, with high-throughput optical links), and service regions will expand to include larger volumes of the South Pole and Far Side. Position, Navigation, and Timing services will expand to more Global Navigation Satellite System (GNSS)–like capabilities by providing services on a global or regional basis. The evolution of lunar comm/nav capabilities will close knowledge gaps to enable an evolution of comm/nav capabilities and concept of operations for Mars missions.

1.3.2.1.2 Habitation Systems

The habitation sub-architecture is a group of capabilities providing controlled environments to ensure crew health and performance over the course of missions. This functionality extends across multiple applications throughout the architecture and is tailored to suit the location and environment (e.g., deep space, lunar surface, Martian surface). Common habitation functions include Environmental Control and Life Support (ECLS), power, communications, thermal control, command and data handling, , extravehicular activity (EVA) support (e.g., ingress/egress, suit services, worksite accommodations, etc.), crew habitability (e.g., hygiene, food and nutrition, waste management, sleep, crew exercise, etc.), crew health (e.g., health and medical care, human performance, psychological support, etc.), and crew survival (e.g. pressurized suits, safe haven, etc.), amongst others. These functions may scale in size and complexity based upon crew size, mission duration, operational environment, and the ability to share functionality through interfaces with other elements (e.g., consumables and power transfer). As such, the volume and structure supporting habitation can vary drastically and potentially include modular, connected, pressurized volumes of various materials (e.g., inflatable soft goods, metallic structure, in situ constructed elements, etc.). While crew size and mission duration are primary factors in scaling the appropriate habitable volumes, other factors such as gravity environment, crew tasks, and required motions (e.g., supportability of on-board equipment; accommodation of science and technology utilization; and logistical stowage and resupply that require controlled, pressurized environments) also factor into overall volume. Some key trades to help scope such habitation elements include EVA ingress/egress methods, logistics resupply needs, and use of regenerable ECLS systems. To maximize the availability of crew time to perform science and technology utilization activities, as well as maintaining nominal operation in each operational environment while uncrewed, it is also critical that habitation elements utilize system autonomy (e.g., vehicle/element control and operation—planning/scheduling/execution, fault management—identification/recovery, robotic assistance, etc.) as well as provide crew control (i.e., manual operations, software override) for critical functions.

1.3.2.1.3 Human Systems

The human systems sub-architecture covers the collective on-board crew, crew support systems (including health and medical, performance, vehicle systems, crew protection, etc.), and mission systems (e.g., operations and facilities) required to execute missions, incorporating any partner support.

The humans who embark on the exploration missions are the most critical component of the campaign to get humans to the Moon and, ultimately, to Mars. Vehicles, systems, training, and operations must be designed around the "human system". In order to provide human-rated systems, standards for design and construction, safety and mission assurance, crew health and performance, flight operations, and system inter-operability are applied.

1.3.2.1.4 Logistics Systems

The logistics systems sub-architecture includes the systems and capabilities needed for packaging, handling, staging, and transferring logistics goods, including equipment, materials, supplies, and consumables needed to support use cases and meet architecture functional needs. This sub-architecture also includes approaches and capabilities for addressing trash and waste management. During the initial part of the campaign, it is anticipated that the capability for logistics goods and consumables will be limited to those that arrive with the crew. As time advances, additional functions are introduced into the architecture. The logistics needs will broaden as the sub-architectures mature. Over time, the architecture will require solutions for increasing mission duration for Mars. The need to deliver elements, payloads, cargo, experiments, and larger quantities of logistics and to better address inventory management, trash, and waste disposal functions necessary to support the missions and meet planetary protection requirements will increase. As the sub-architecture matures, the capabilities can continue to grow to take advantage of increased automation and/or in situ resource sourcing of logistics to support increased mission durations.

1.3.2.1.5 Mobility Systems

The mobility sub-architecture is a group of capabilities and functions that enable the robotic-assisted mobility of crew and/or cargo on and around the destination, including EVA systems. This sub-architecture group extends the range of exploration and external operations in support of exploration and science. It spans robotic and crewed systems with both pressurized and unpressurized capabilities depending on the use-case functions required. Mobility systems will likely need to interface with other sub-architecture capabilities like power, CPNT, habitation, and logistics transfer.

1.3.2.1.6 Power Systems

The power sub-architecture is a group of capabilities that support the function of providing electrical energy to architectural elements. These capabilities include components and hardware for power generation (e.g., solar arrays, Fission Surface Power [FSP]), power distribution (e.g., electrical cables, induction), and energy storage (e.g., batteries, regenerative fuel cells). A primary aspect of the power sub-architecture is interoperability, including standardized power interfaces (either hard or inductive connections) and compatible power quality standards. The power sub-architecture will include the coordination of missions where elements are expected to provide their own power to the development of any needed energy infrastructure to support future needs.

1.3.2.1.7 Transportation Systems

The transportation sub-architecture is the collection of capabilities that provide the transportation functions for all phases of the Moon and Mars missions for both crew and cargo, including in-space; Entry, Descent, and Landing (EDL); and ascent for all Earth, Moon, and Mars phases. The transportation systems will need to interface with or be incorporated into a variety of systems and payloads, including habitation, and other human support systems, as well as refueling or recharging systems, all in diverse environments including in-space and surface conditions. Initial lunar segments will include transportation capabilities for the transit of crew and cargo to cislunar space, the landing of crew and cargo on the surface, crew and limited cargo ascent to cislunar space and return to Earth. As the architecture expands toward Mars, the transportation sub-architecture will evolve to include Mars transit, EDL, and ascent systems for cargo and crew.

1.3.2.1.8 Utilization Systems

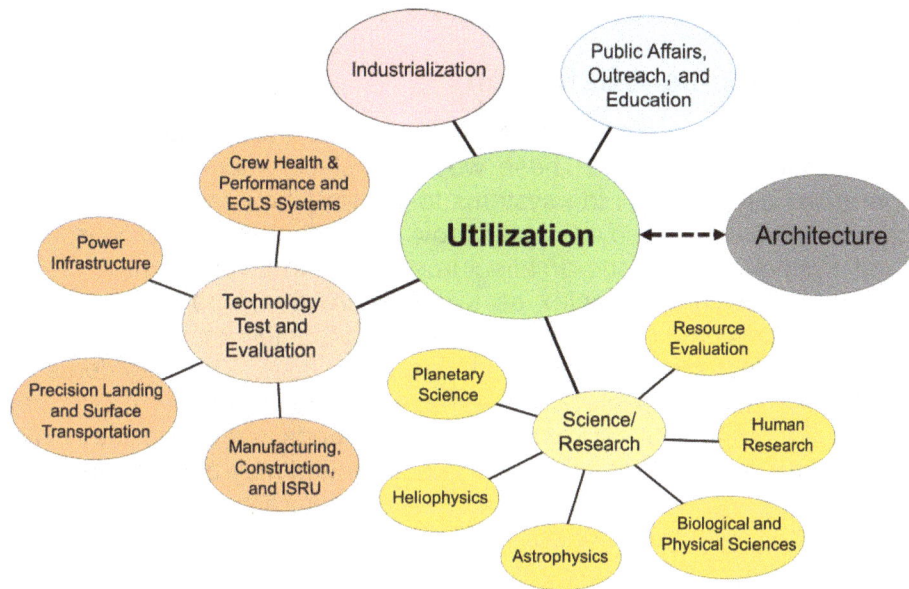

Figure 1-6. Visualization of Utilization Areas

The M2M Strategy document[4] defines Utilization as the "use of the platform, campaign and/or mission to conduct science, research, test and evaluation, public outreach, education, and industrialization." In this document, the term "utilization" is used generically to encompass all areas of utilization; specific terms, such as "science or technology demonstration," are used where the meaning is more specific. The utilization systems sub-architecture is a group of capabilities whose primary function is to accomplish these science, technology, and similar activities, including sample and utilization cargo return to Earth. In this sense, the M2M architecture provides a platform of functions to a broad set of organizations in support of their needs. Inherent in the Moon-to-Mars architecture is that all of the sub-architectures ultimately support utilization. Support for utilization systems will levy functions and use cases on all other sub-architectures. The major utilization areas of emphasis for the M2M Architecture are depicted in Figure 1-6.

Utilization is achieved through not just the capabilities in the utilization systems sub-architecture, but the entire architecture. For instance, a technology may be demonstrated under the umbrella of utilization on one mission and, through technology maturation, provide essential services as part of the exploration platform on subsequent missions. Similarly, some items may serve multiple functions, e.g., multi-purpose cameras used for both science and operations, as well as equipment shared between human research and medical operations. However, systems whose primary purpose is to achieve utilization, and not just enable, will be included in the utilization sub-architecture.

1.3.2.1.9 Future Sub-Architecture Development

As the focus of the Architecture Framework is to establish the process for recurrent architecture definition and refinement, so will the sub-architectures evolve. As described above, the initial sub-architectures were established based on knowledge gained driving system requirements. However, it is expected that additional sub-architectures will be beneficial to support systems and allocations in future revisions. Candidate sub-architectures, such as Command & Data Handling,

---

[4] NASA's Moon-to-Mars Strategy and Objectives Development. NP-2023-03-3115-HQ.

and Construction or In-Situ Resource Utilization (ISRU) functions, will be assessed and scoped during upcoming Strategic Analysis Cycles and/or ACRs.

1.3.2.2 Campaign Segment Definition

The purpose of segments is to capture at a phase in time the interaction, relationships, and connections of the sub-architectures. These would most commonly be typified by reference missions or operations use cases of the systems to illustrate how systems will work together to achieve objective satisfaction. These examples provide the context for the allocation of functions to elements and systems in the sub-architectures rather than prescriptive solutions. These segments will grow increasingly complex as systems are developed and added to the sub-architectures. The segments are crafted in a manner such that the knowledge gathered earlier in the campaign informs the implementation of the latter part. The segments integrate the exploration, utilization, and sustained development of the Moon with preparation for the exploration of Mars. The segments will serve to integrate needs and capabilities over time but are not a defined launch manifest as systems from a later segment may begin to appear as available. Further, in representing the context of the sub-architecture interactions, segments are not limiting the types of missions that may be designed and flown. As systems are built, novel operations and uses are expected.

The segments, described in detail in Section 3.0, reflect the current Moon-to-Mars effort and provide open opportunity to refine and include use cases as systems and technologies mature to infuse into the architecture. The segments and their content will evolve through the annual ACR cycles to reflect the inputs, capabilities, and needs identified across the partners to achieve the M2M Strategy.

1.3.3 Architecture Definition Process

Having established the necessary components to decompose objectives "Architecting from the Right" and the framework to correlate the systems "Executing from the Left," the process by which these will be integrated remains. As acknowledged, the process is established to enable an iterative allocation to programs and projects and infusion of solutions, technologies, and capabilities that emerge over time to address the strategy objectives. This process is managed by NASA's Exploration Systems Development Mission Directorate through the coordination of Strategic Analysis Cycles (SAC). These cycles will occur annually to prioritize the work and studies needed to address open questions, identify potential architectural drivers to buy down mission risk, coordinate with partners, and identify and resolve gaps in the architecture. The cycles will conclude with study findings and/or updates and iteration to the Architecture Definition Document and supporting products to be reviewed at the annual Architecture Concept Reviews in support of the NASA planning process.

As an iterative process, these cycles will need to both enable the definition of new elements or systems as they are added to the architecture by defining the allocated functions and needs, and also update and modify the architecture as existing elements and programs mature. Additionally, the SAC process will need to include assessments or studies for how emerging technologies or solutions identified, whether within NASA or from partners, could address architecture needs or modify the future segments if realized. This complex analysis process will reflect a diversity of viewpoints, perspectives, and ideas from stakeholders and partners.

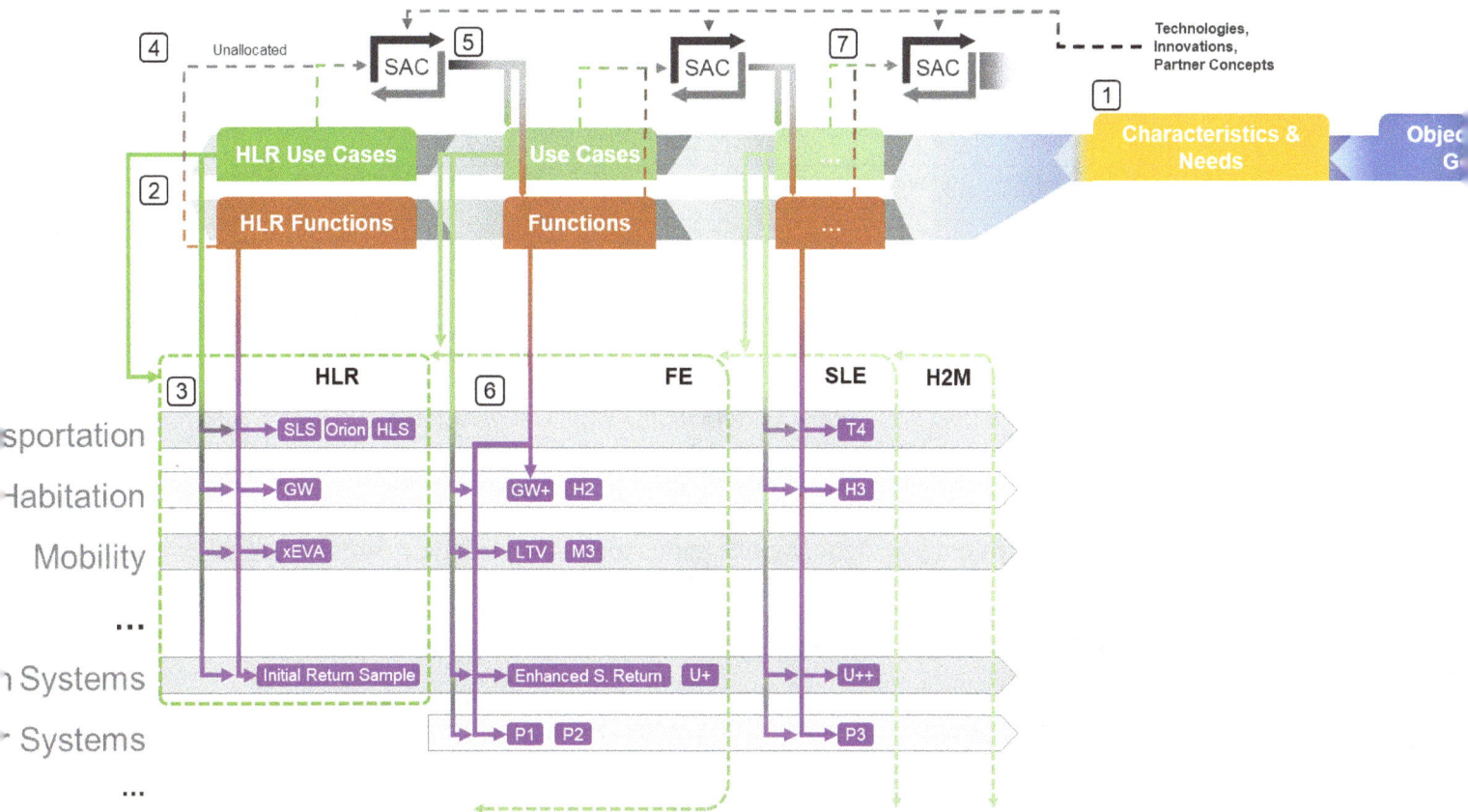

Figure 1-7. Illustration of the Architecture Definition Process

Table 1-3. Iterative Architecture Process Steps

1	Objectives decomposed to Use Cases and Functions
2	Element allocations and traceability performed to segments
3	Program requirements and Concepts of Operation implementation allocated to architecture needs
4	Unallocated functions (gaps) re-enter SAC process w/ partner inputs/concepts
5	SAC trades and analysis identify element solutions or definition of new program/projects
6	Definition of next segment and included elements begins
7	Repeat

The architecture definition process is shown in Figure 1-7 and reflects the intersection of the Right and Left principles outlined in the M2M Strategy. Examples and representative systems using known sub-architectures, segments (discussed further in Section 3.0) and Elements are used to illustrate this iterative process. This process reflects the reality the systems, functions, and needs of the most immediate segment are known and that significantly fewer allocations are made as the segments process to the right. Systems reflected in the current programs and projects are already executing their development and, in some cases, have conducted their first flights, such as the Space Launch System and Orion. Modifications to these existing systems should be limited or carefully traded in future segments. The SAC process will need to consider the programmatic trades in any allocation, whether existing or new systems are used, for cost, schedule, technical, and risk factors. The process steps highlighted in Figure 1-7 are outlined in Table 1-3.

The SAC trade studies will continue to evaluate concepts and analysis to identify possible solutions to address unallocated functions and potential alternatives. Coordination with both internal NASA and external partner communities will be a key enabler to identify solutions that can most effectively address objective satisfaction. Inputs of technological advancements, alternate concepts, and other innovations can be assessed for satisfaction to meet the integrated architecture needs during the Strategic Analysis Cycles. These assessments will mature and refine allocations in partnership with the executing element or partner leadership to ensure traceability from the use cases and functions into the requirements and concept of operations that formally establish the design process for execution. The SAC process will also consider technology advancements, alternative solutions, and different concepts to identify efficiencies or priorities for development in future segments. These efforts will inform how future systems and elements are instantiated and developed as systems mature.

2.0 ARCHITECTURE DECOMPOSITION

A similar systems engineering process applied at the strategic level in Section 1.0 can be used as a framework for the architecture by addressing the six key questions: Who, What, Where, When, How, and Why? (Figure 2-1.) Different stakeholders may find the answer to one of these questions more compelling than others: for example, engineers tend to focus on "How?," whereas technology developers may be more interested in "When?"; partners want to know "Who?," and scientists may be keen to discuss "Where?" and "What?" To reach consensus and move forward, an exploration architecture must address all six questions, but reiteration and negotiation may be required. The answer to any one question is less important than ensuring that the answers to all six fit together as an integrated whole.

Figure 2-1. Elements of a Compelling Architecture Story

2.1　EXPLORATION STRATEGY—"WHY EXPLORE?"

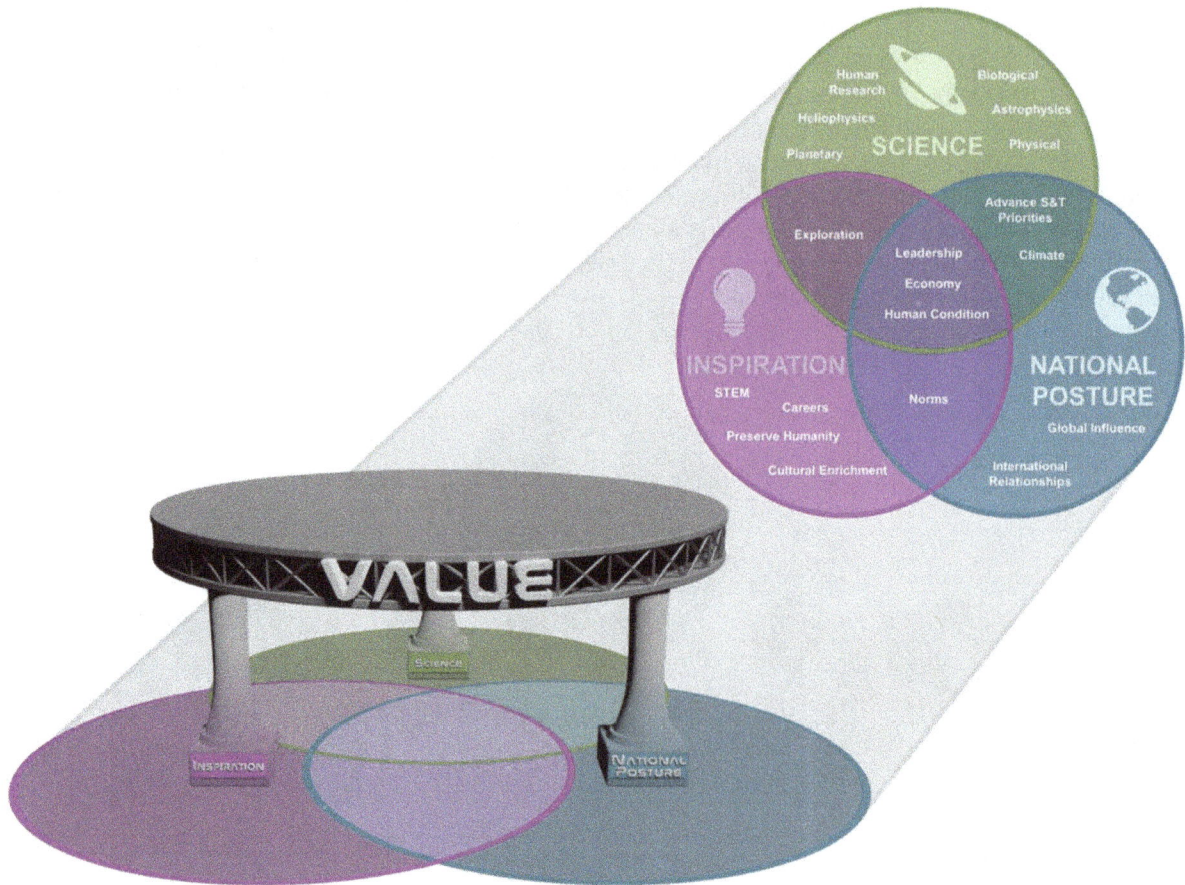

Figure 2-2. Three Pillars of Exploration from NASA's Moon-to-Mars Strategy and Objectives Development Document[5]

Systems engineering is predicated on the motivation, which is the fundamental goal. Why do this? For the blueprint vision and Moon to Mars endeavor, along with its goals, objectives and subsequent architectural wireframe, the question is: Why send humans into space? Creating a blueprint for sustained human presence and exploration throughout the solar system provides a value proposition for humanity that is rooted across three balanced pillars: science, inspiration, and national posture. Each pillar contains both unique and intersecting stakeholder values that together form the value proposition for the blueprint vision, starting with the Moon to Mars endeavor (shown in Figure 2-2). While different individuals identify with different values, it's NASA's responsibility as a steward of taxpayer dollars to consider the entire landscape of motivating factors that underscore our society's answer to Why Go? Uniquely, by balancing all the factors, NASA positions the Moon to Mars strategy for longevity and success: It is not subject to whims or leadership overhauls. Instead, it is rooted deeply in a broadly relevant, largely unchanging value system. So, Why Go? It's these three pillars, combined and with their intersections, that are why humans go into space, as illustrated in Figure 2-2.

---

[5] NASA's Moon-to-Mars Strategy and Objectives Development. NP-2023-03-3115-HQ.

2.1.1 Science

The pursuit of scientific knowledge – exploring and understanding the universe – is integral to the human space exploration endeavor. Just as the James Webb Space Telescope informs about the history of time, answers gained on the Moon and Mars will build knowledge about the formation and evolution of the solar system and, more specifically, the Earth. From geology to solar, biological, and fundamental physics phenomena, exploration teaches about the earliest solar system environment: whether and how the bombardments of nascent worlds influenced the emergence of life; how the Earth and Moon formed and evolved; and how volatiles (e.g., water) and other potential resources were distributed and transported throughout the solar system. Space exploration teaches about human and plant physiology in extreme environments, how to mitigate engineering and health risks, as well as how to perform complex operations in harsh planetary environments. Space provides a unique vantage point to greatly amplify current learning on Earth. Biological and physical systems can be observed in reduced gravity, bringing out second and third order effects that are otherwise overwhelmed in the gravity environment. The history of our Sun is preserved in lunar soil, examination of which enables solar activity predictions and space weather forecasts, which in turn supports lunar and Martian exploration. Specific frequency ranges available for use only in space (due to interference by other Earth-based signals or the atmosphere) allow probing the deepest space and time of the universe. While remote sensing is a great aid, robotics and direct human engagement with and visitation of other bodies in the solar system ultimately reaps more data more effectively.

2.1.2 National Posture

By its very nature, achieving a vision of space exploration establishes national strength in science and technology innovation and competitiveness, which supports economic growth and global position. Hard technology problems solved in space have far-reaching implications for other Earth-based challenges and industries, and in many cases, spin off their own disciplines. For example, the term "software engineering" was crafted for the development of the guidance and navigation systems on Apollo spacecraft. Food safety standards and telemedicine likewise originated with NASA in an effort to enable longer duration human space flight. NASA technology, spin-offs, and investments fuel growth in American industry and support quality, high-paying jobs across the country. Specifically, NASA's contracts and partnership with domestic commercial space has resulted in $15 billion in private investments in space start-up companies in a single year of this new era alone, with the majority of those investments in United States companies. Commercial space activity impacts other industries such as agriculture, maritime, energy, and homeland security, producing ripple effects throughout the economy. Additionally, because there are no geographic bounds in space, exploration lends itself to international partnerships to achieve feats that might not otherwise be possible. Bolstering international partnerships, economic competitiveness, and global influence likewise reinforces national security interests.

2.1.3 Inspiration

The "Moonshots" of the Apollo Program became a metaphor for how we as a nation could take on an audacious challenge and succeed through hard work and determination. The "Moonshot" metaphor has since been applied to inspiring and seemingly insurmountable challenges from curing cancer to developing fusion power. Apollo inspired a new generation of engineers and scientists in education and career pursuits supporting visionary work. The International Space Station and other space partnerships model how people from many nations can live and work together toward a common purpose. These next steps in space exploration can likewise inspire an all-new generation – the Artemis Generation – in science, technology, engineering and mathematics studies that support the great enterprises of voyaging into space and overcoming the most difficult challenges currently faced on Earth.

2.2 LUNAR ARCHITECTURE STRATEGIC ASSESSMENTS

The effort to return humans to the Moon has been addressed at a strategic level first by answering the "Why," as documented in NASA's Moon-to-Mars Strategy and Objectives Development document. This strategic plan ensures that the lunar architecture must consider a range of stakeholder needs, including the long-term goal to enable Mars and other deep space exploration. Definition of the architecture and the methodology to achieve it is fundamental to the leadership needs reflected in the "Why." By implementing an architecture that can be responsive to innovation and developments and inclusive of partners, the endeavor will enable benefits reflected in terms of both the economy and the human condition. Working from both the Blueprint Objectives and the array of available Mars studies, several key characteristics of the lunar architecture have been derived. Throughout all of these decisions, the Responsible Use (RT-6) Tenet is applied to ensure consistent application of policy, legal, and ethical frameworks. If areas of uncertainty in how policy or standards should be applied to the objectives or architecture are identified, they will be elevated for resolution to Agency leadership.

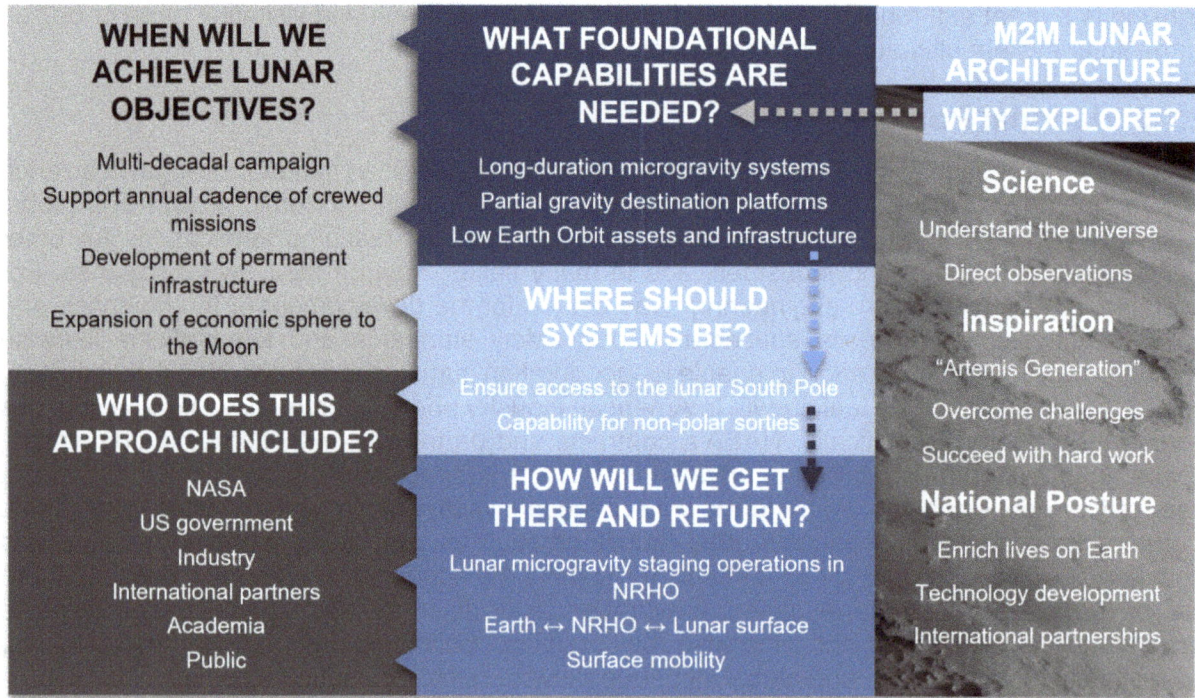

Figure 2-3. Lunar Architecture Decision Flow Starting with "Why?"

As illustrated in Figure 2-3, answering the "Why" for lunar exploration is only the first step in the decision process. Answering, or exploring the option space for, other big architecture questions ("Where," "What," "Who," "How," and "When") helps define the key characteristics of the lunar exploration campaign.

2.2.1 Key Lunar Decision Drivers

2.2.1.1 "What" Foundational Capabilities are Needed?

As decomposition of the Objectives, captured in the Strategy and Objectives document above, indicates, several technological, scientific, or human condition insights are needed to inform Mars architectural decisions. These multi-dimensional objectives across the science, technology, and infrastructure development goals will need to be supported by foundational platforms from with the crew will operate. These systems will provide the crew support to retrieve and return samples, deploy instrumentation or technology demonstration, research in situ resource utilization, understand the human condition in long-term deep space exploration, and much more. The ability to support and the decomposition of the capabilities needed to accomplish Science and Infrastructure Objectives will be key characteristics for cislunar and surface destination systems. Common across all architectural studies to date is the need to provide demonstration and test environments across dynamic space weather conditions, deep space microgravity, partial gravity, and transitions between them. These environmental drivers must be paired with increasing representative operational durations to establish sufficient design, engineering, and demonstration drivers in the architectural approach.

From these assessments, two key destination systems and the ability to transition between them are derived as platforms for this development. First, the ability to stage long-duration microgravity systems in the deep space or near–deep space–equivalent environmental conditions provide analogous drivers to the transit conditions to and from Mars. This platform will necessarily need to provide the human habitation support and ability to function with reductions in crew-managed reliability, maintenance, and ground intervention associated with near-Earth systems (RT-5 Maintainability and Reuse). Further characteristics beneficial for any microgravity platform are the ability to aggregate elements autonomously and to support incremental build-up to prepare for the eventual accumulation of systems necessary to achieve Mars capable transit systems.

Second, as a destination platform, it is necessary to provide systems deployment and aggregation in partial gravity with what can be considered a "hostile" atmospheric or, in the lunar case, no atmospheric environment for human systems. This surface platform as an aggregation of elements will also provide the opportunity to demonstrate the necessary components for achieving Mars-forward systems for human-conducted surface exploration. These systems will need to support increasingly longer crewed exploration periods and be expandable to accommodate the breadth of the objectives laid out towards the Mars goal.

Additionally, the architectural approach will leverage available low-earth-orbit assets and infrastructure to support the lunar and Mars objective accomplishment. In this regard, exercising objective capabilities at the lowest energy state (whereas performance demand can be considered proportional to the resources and/or programmatic needs) will be applied throughout (RT-8 Leverage Low Earth Orbit).

2.2.1.2 "Where" Should the Systems Be?

From the definition of the "What," it is necessary to support the architectural approach in the microgravity and surface platforms; "Where," in relation to the Moon, became the next systems engineering driver. To assure the platforms' support of long-term objectives and the balanced-systems approach of the "Why," numerous studies of lunar system locations have been conducted by NASA and others. The primary consideration of "Where" is to ensure surface system access and optimization to the lunar South Pole; however, this ensures access to non-polar locations as well.

With an eye to the engineering for systems demand, the lunar South Pole has several key driving characteristics to enable systems development for the Moon-to-Mars Architecture. First, from a flight performance perspective, the lunar South Pole provides a bounding condition for vehicle translation or delta-velocity costs. These performance drivers are one of, if not the, most significant condition in transportation system design. Vehicles and reference missions designed to achieve landing at the South Pole can provide future flexibility to reach global locations through planning and certification. This is unlike the Apollo vehicles and systems, which, when directed to reach the Moon at essentially the earliest possible time (answering the "When" question first), necessarily selected the "easiest" lunar landing sites on Earth-facing, near-equatorial regions and lacked the performance and systems capabilities for global and/or polar landings, would they subsequently have been desired.

Second, lunar poles have considerable characteristics that support multiple scientific and engineering values, and the South Pole provides more opportunity for these conditions in designing systems to ensure extensibility to other lunar locations and future Mars needs. One key enabling characteristic of the South Pole is the lighting conditions. At the lunar equator, the solar illumination occurs in 14 days of continuous daylight and 14 days of continuous darkness. These are the lunar cycles we are so familiar with as viewed from Earth. However, at the South Pole, the Sun is seen very low on the horizon as during the extreme summer nights at Earth's poles. Unlike Earth, the extreme terrain on the lunar South Pole provides significant variation, resulting in "peaks" of light that can provide lit conditions for much of the year and "valleys" of darkness that never see the Sun. These peaks or ridges along craters provide advantageous locations to stage systems for longer-duration operations. However, while advantageous for illumination, the peaks and ridges are more challenging for navigation and placement of elements. These factors must be considered in the architectural approach and element designs. This lighting environment will be an enabling feature of the polar region to represent Mars-forward precursor missions and aggregation of surface elements for longer-duration test and demonstration.

Finally, the lighting conditions in the south polar region also contribute to unique scientific opportunities. Although the lunar surface was found to be void of volatiles as they are stripped away by the solar wind, sites of permanent darkness in the polar regions could preserve volatiles collected throughout the Moon's past. This region is among the oldest parts of the Moon, older than any explored during Apollo. The volatiles, likely trapped as ice, could reveal valuable knowledge about the history of the inner solar system, including when life gained a foothold on Earth. Just as ancient ices hold scientific value, lunar samples from this area will increase our knowledge of the history of the Moon itself. Additionally, these ices could serve as valuable resources for use during future exploration. Finally, the peaks of light at the South Pole are an enabling characteristic to support extended durations of human-tended surface operations to provide the infrastructure capabilities for sustainable and lasting development and research.

In both these environments, several architectural drivers result. In particular, the systems of power and thermal to operate in environmental extremes, provide surface mobility, and allow the aggregation of infrastructure are possible. Given the necessary development of platforms and systems at these locations over time, the application of interoperability and commonality will be a key enabling characteristic (RT-7). The ability to deploy, upgrade, and develop systems across the platforms will be critical to the evolution and continued operations of the integrated architecture. These combined reasons ensure that the South Pole is a significant feature in the Moon-to-Mars Architecture definition, while also maintaining and supporting the ability to visit diverse non-polar sites.

2.2.1.3 "How" Will We Get There and Return?

An understanding of the driving surface of the South Pole destination for long-term infrastructure, the need for global periodic access, and the development of a long-duration cislunar platform, inform the architectural driver of "How" to place the lunar microgravity staging operations. From a variety of studies and alternatives, the Near Rectilinear Halo Orbit (NRHO) will be utilized by the lunar architecture. This orbit meets the needs of several key characteristics, including the long-duration staging through minimal propellent demand for orbit maintenance, accessibility to the lunar South Pole and other global access on a frequent and recurring basis, and consistent access for crew and cargo to and from Earth while still providing near–deep space environmental conditions with near-continuous illumination and limited lunar albedo (or reflectivity of light and heat) to the orbiting platform.

Having established the NRHO architectural orbit, the ability to transport crew, cargo, and support systems to and from the destinations can be decomposed. These systems are driven by the sizing performance splits across the architectural destinations to traverse the regions from Earth to cislunar space to the surface. Crewed transportation systems will be driven by the need to launch, transport, and safety mitigate potential contingencies and risks in two key transportation regimes: first, crew accessibility to and from Earth to NRHO platforms, and second, to and from NRHO to the surface destinations to support either South Pole or non-polar mission selection. The crew transportation access, in conjunction with destination systems, will necessarily need to ensure the safety and responsive planning for Crew Return (RT-3) for potential contingency scenarios.

Transportation objectives are some of the earliest objectives in the architecture and most established systems, given that they are necessary first steps to the human return to the Moon, enabling subsequent objectives. These systems are developed with several key characteristics applied, including the ability to achieve missions with sufficient frequency and opportunity. With the scope of objectives and highly coupled architectural aggregation approach, the ability to ensure timely and consistent launch and mission opportunities is a key characteristic. In addition to crew transportation, systems will also need to support the launch and delivery of cargo across a range of mass and volumes to support the element aggregation, logistics, and maintainability across the architectural lifetime and destinations (RT-5 Maintainability and Reuse). Systems capable of re-use will have significant benefit, not just for transportation but across the architecture, to reduce launches and continue to enable long-term objectives. Given the significant considerations in transportation objectives, the needs to support docking, deployment, and disposal, when required, with the minimum involvement of crew intervention will be key. The ability to maximize Crew Time (RT-4) to achieve utilization and other objectives would be compromised as crew time is increased to support routine operations, maintenance, and services.

Although the largest and most recognizable transportation systems are those that carry the crew through space, the ability to support mobile operations on the lunar surface is a key characteristic to achieve many of the Blueprint Objectives. Mobility systems on the lunar surface are necessary to enable the myriad scope of objectives that must be accomplished at points across the surface that would be impractical to reach or inaccessible to crew traveling only on foot from the landing location. To maximize crew time (RT-4) applied to the utilization objectives, the ability to transport them efficiently and safely between desired surface locations is paramount. This capability is similarly needed for future Mars exploration and is essential to the exploration plans to enable the crew to travel to increasingly farther points from the landing sites, explore regions for which landing is not feasible, and carry and transport samples or utilization payloads.

To enable these complex operations, the support of a robust, secure communication, position, navigation, and timing system will also be critical. The volume of data to safely monitor, command, and control active vehicles, both crewed and uncrewed, will be a key characteristic of the integrated architecture. The number of systems in both cislunar and surface operations will also generate the need to handle multiple simultaneous streams of data and telemetry. Management of these systems and functions across the distributed architecture through interoperable and expandable systems will be a key characteristic to accomplish lunar objectives.

2.2.1.4 "When" Will We Achieve Lunar Objectives?

As the lunar campaign has already begun, the key characteristics to address the "When" question are more appropriately addressed as the time frequency, or how often. Driving the systems to support an annual cadence of crewed lunar missions is a need that flows throughout the system development from ground processing and launch facilities, to development and assembly timelines, to the assets necessary to support those missions. Turnaround and processing times will additionally be a key characteristic for any system reuse driven by the transportation objectives. Further, the demand for logistics supply, repurpose, and disposal will be key considerations in the opportunity frequency for the architecture. Logistics demand is a significant derived capability necessary to support increasing mission durations to accomplish the Blueprint Objectives at both cislunar and surface platforms. Periods between crew flights will drive characteristics for the assets to provide ongoing value and benefit during tele-robotic or autonomous operations.

These opportunities, representing a diverse suite of frequent crewed and uncrewed operations through permanent infrastructure, will provide significant opportunity for commercial and space development. The Agency objectives will have the opportunity to be addressed through a variety of approaches, innovations, and partners. One of the key Recurring Tenets applicable to addressing "When" is RT-9, the Commerce and Space Development. As has been demonstrated in low-earth-orbit (LEO) with the development of commercial systems and the partnerships that result, the ability to foster the expansion of the economic sphere to the Moon will enable the support of U.S. industry and innovation. Creative solutions to meet multi-user needs, responsiveness to opportunities, and the shared support of lunar exploration across industry and partners will be necessary to support sustainability and durability of the architecture.

The planned campaign spans a multi-decade period, establishing permanent footholds in cislunar space and on the lunar surface, developing and deploying major human-rated transportation systems to the Moon and Mars, and developing and deploying lunar and Martian surface infrastructure to enable humans to live and explore once they arrive. The term "sustainable" can have different meanings, depending on the context. For the exploration campaign, several definitions apply. Financial sustainability is the ability to execute a program of work within spending levels that are realistic, managed effectively, and likely to be available. Technical sustainability requires that operations be conducted repeatedly at acceptable levels of risk. Proper management of the inherent risks of deep space exploration is the key to making those risks "acceptable." Finally, policy sustainability means that the program's financial and technical factors are supportive of long-term national interests, broadly and consistently, over time.

2.2.1.5 "Who" Does This Approach Include?

Having established all the component parts of the architecture, sizing for systems is designated to include up to four crew during an integrated mission. These crew members will thus be enabled to conduct the scientific, technological, and developmental objectives for which the human mind is most suited. Again, maximizing the time of the crew members to support these objectives is a Recurring Tenet (RT-4) in the architectural selection and decomposition. The ability to support four crew members provides the opportunity to assign various tasks ranging from piloting to utilization and operations among the participants. The crew operations will provide a gradual build-up approach to demonstrate the technologies and operations necessary to live and work on planetary surfaces and extended deep space microgravity environments, including a safe return to Earth. Risk is inherent in any type of space flight; however, it is a key consideration in the context of human space flight. As one of the Recurring Tenets, Crew Return (RT-3) is a key characteristic across all architectural domains. The application of risk management, fault tolerance, and integrated human-rating certification is necessary at the architectural level. Given the necessary infusion of this approach into all systems, contingency, aborts, and risk management are treated as an applied characteristic across the architecture.

The humans who embark on the exploration missions are the most critical component of the campaign to get humans to the Moon, and ultimately to Mars. Vehicles, systems, training, and operations must be designed, developed, and certified to be safe and reliable for, compatible with, and in support of the "human system" as an integrated system to accomplish the mission with an acceptable level of human risk. Human-rating is the process of designing, evaluating, and assuring that the total system can safely conduct the required human missions, as well as incorporating design features and capabilities that both accommodate human interaction with the system to enhance overall safety and mission success and enable safe recovery of the crew from hazardous situations. As part of human-rating, standards for design and construction, safety and mission assurance, health and medical concerns, flight operations, and system interoperability are applied. Human-rating is an integral part of all program activities throughout the life cycle of the system, including (but not limited to) design and development; test and verification; program management and control; flight readiness certification; mission operations; sustaining engineering; and maintenance, upgrades, disposal, and ground processing. NASA will lead/integrate the distributed team of government, commercial, and international partners that develop and implement hardware, software, and operations supporting exploration. Both nominal and contingency scenarios must be part of the overall development of the mission, hardware, software, and operations to arrive at a reasonable level of risk. The crew will require many months/years of Earth training across numerous vehicles and systems in a compressed timeframe to prepare for the mission.

In addition to the crew themselves, the development of the myriad of systems, operations, and capabilities to meet objectives will require the support of industry and international collaborations (RT-1 and RT-2). The Moon-to-Mars architecture approach is to enable a variety of support mechanisms and contributions to enable innovation, economic development, and the inspiration foundation to address Why We Explore (Section 2.1). Characteristics to enable this include architectural robustness to infuse innovative solutions and technological advancements over time. The iterative methodology, flexibility in design solutions, and ability to perform responsive mission planning for future developments will be key considerations in the effectiveness of the architecture to meet the objectives.

NASA has a long, successful history of working with a diverse community of international partners to advance common space exploration and science objectives. NASA is committed to building on and broadening these global partnerships as part of the Moon-to-Mars objectives. NASA has numerous international partnerships already in place and is engaging in ongoing dialogues bilaterally and multilaterally with international space agencies from around the world to identify new, mutually beneficial opportunities for collaboration.

Building upon the ISS/LEO experiences of more than two decades, operational flexibility to demonstrate the capability to integrate the multi-party contributions, aggregations of systems over time, and increasing complexity will be needed to address long-term Mars-forward development. The coordination of integrated ground, launch, and flight systems through both crewed and uncrewed regimes and multiple planetary bodies will require a significant leap forward in the complexity of mission operations from the existing experience base.

2.2.2 Unique Considerations for the Moon

Although NASA has conducted human exploration on the lunar surface previously in the Apollo program, there are still unique aspects to consider for the current lunar architecture. With the desire to seamlessly expand to long-term, sustainable exploration while preparing for human Mars exploration, the M2M architecture must remain flexible to plan for the future campaign with current programs and elements in development, adjust to the actual flight systems as the elements mature and are deployed, and accommodate new contributions. This allows for an incremental increase in capabilities for lunar exploration, gradually building up functionality to achieve the Agency objectives.

The most recent human space flight exploration, and the majority of human space flight hours of experience, has been conducted in LEO. There are several major differences in concept of operations between LEO and cislunar missions. For one example, abort capabilities back to Earth vary in duration. With exploration interest in lunar South Pole locations and a cislunar platform in NRHO, aborts back to Earth are more complex and take days rather than hours, as is the case from the International Space Station (ISS). These durations significantly complicate or eliminate crew rescue options that may be available in LEO. In another example, crew will transition between micro- and partial-gravity environments, eventually doing so after extended durations in microgravity without the support that crew members experience upon their return to Earth after long missions on the ISS. Testing out the concept of operations for surface exploration with deconditioned crew is one aspect that will also help prepare for Mars exploration. Further, the unique aspects of the South Pole, in term of lighting, terrain, and other environmental considerations, present unique challenges to the missions and strategic planning. These include the relatively constrained area of the South Pole, which is advantageous, not only to NASA, but also to other commercial, scientific, international, or other lunar exploration plans.

2.3 MARS ARCHITECTURE STRATEGIC ASSESSMENTS

In the five decades since Dr. Wernher von Braun proposed NASA's first human Mars architecture, NASA has pivoted from one exploration point design concept to another, many optimized around heritage programs or emerging technologies of particular interest. Indeed, half a century of architecture studies have filled our libraries with myriad architecture concepts, all having one thing in common: none of these concepts found traction with stakeholders, many of whom had competing perspectives or needs. The Agency's new Moon-to-Mars Objectives provide a comprehensive framework to ensure that human Mars architectures will meet—or can evolve to meet—more stakeholder needs. After mapping objectives to the required functional capabilities, the architecture team will coordinate with technology and element concept developers and identify the key architecture decisions that must be made. Because decisions in one part of the architecture will ripple through other parts of the architecture, it is critically important that decision makers understand the effect of each decision on the integrated architecture, including differences depending on which decisions are made first. The strategic assessment and campaign segment description described in this document form the foundation for this Mars decision road-mapping process. Later revisions will document Mars architecture decisions as they are made.

To build a compelling architecture that will gain traction with stakeholders, a similar systems engineering process applied at the strategic level can be used as a framework for the architecture by addressing the six key questions: "Who," "What," "Where," "When," "Why," and "How"? To reach consensus and move forward, an exploration architecture must address all six questions, but reiteration and negotiation may be required. The answer to any one question is less important than ensuring that the answers to all six fit together as an integrated whole.

2.3.1 Key Mars Decision Drivers

As noted in at the beginning of this section, the human Mars exploration architecture can be described as a six-sided trade space, shaped by the answers to six key questions: "Who," "What," "Where," "When," "How," and "Why"? (As shown in Figure 2-1.) In laying out the Agency's architecture decision roadmap, it is critically important for decision-makers to understand how these key drivers relate to each other and how the architecture can vary depending on the order in which these decisions are made.

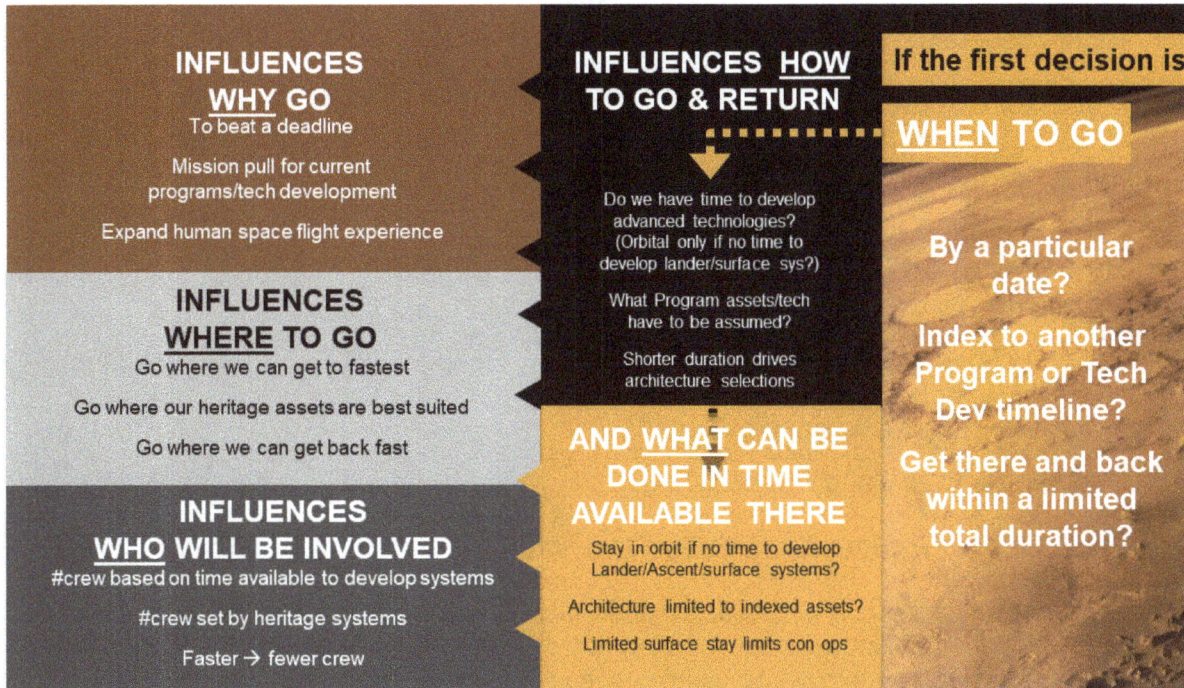

Figure 2-4. Mars Architecture Decision Flow Starting with "When?"

The Apollo program was famously characterized by the mandate of "landing a man on the Moon and returning him safely to Earth before the end of the decade". This prioritized "When?" (within the decade) over other considerations. NASA successfully achieved the goal, but because the resulting architecture was optimized to meet a tight implementation schedule, it was not a particularly extensible architecture, with implications to sustained human exploration of the Moon.

The Apollo program serves as a cautionary tale for Mars: if decision-makers focus on "When?" as an anchoring decision (Figure 2-4), and the answer is a date that does not give us enough time to develop new technologies, then the answer to "How?" would default to heritage or heritage-derived systems. If the specified date is too soon to develop and certify both a new transportation system and new descent, ascent, and surface systems, then the schedule compromise may be an orbital-only or fly-by first mission, followed by surface missions in later years. This affects not just "How?" but cascades to "What?" and "Why?" If instead of a particular date, "When?" is indexed to another event—for example, the timeline of a particular technology development or an Agency funding profile—then certain technologies or assets from other programs may be prescribed, again influencing both "How?" and "What?" If the answer to "When?" specifies both a "boots on Mars" date and a "boots back on Earth" date (in other words, a total crewed mission duration) that will define whether we require new high-tech, high-energy transportation systems capable of shorter mission durations. As shown in Figure 2-4, starting with

"When?" potentially makes the answers to "Why?," "Where?," and "Who?" reliant on the answers to "How?" and "What?".

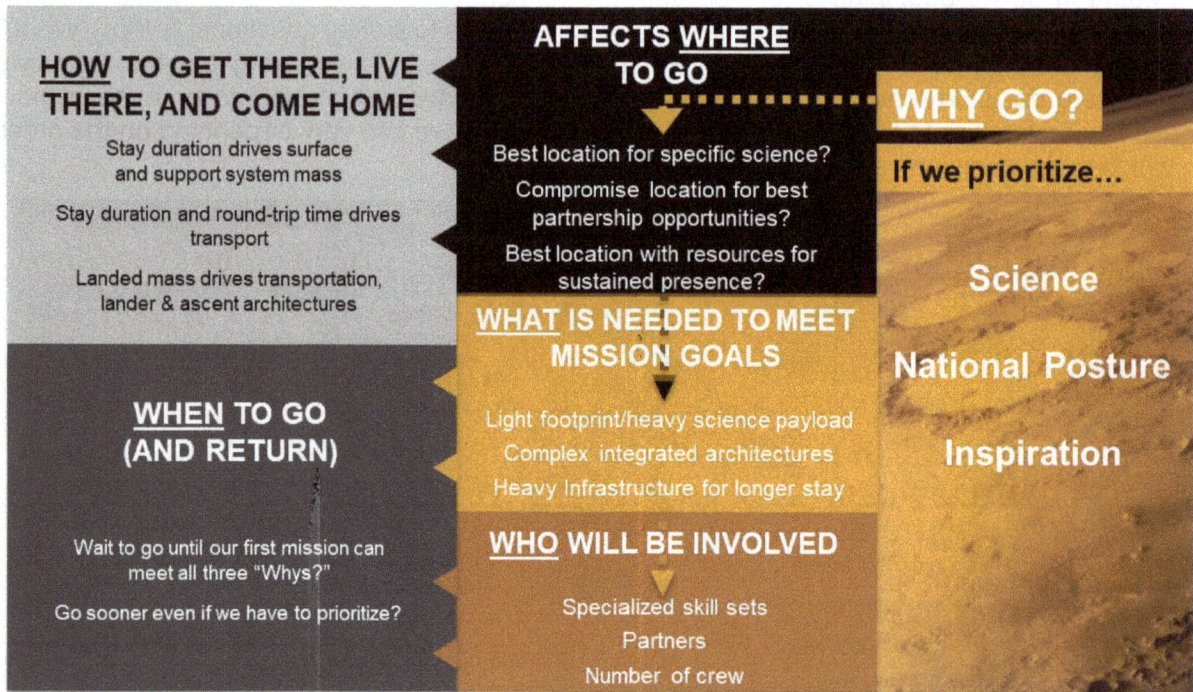

Figure 2-5. Mars Architecture Decision Flow Starting with "Why?"

With few architecture decisions mandated thus far, human Mars exploration offers a unique opportunity to take an objectives-based exploration architecture development approach. NASA's M2M Strategy provides such a framework. In contrast to a capabilities-based approach, an objectives-based approach focuses on the big picture, the "What?" and "Why?" of what NASA should be doing in terms of deep space exploration before prescribing the "When?" or "How?"

As shown in Figure 2-5, NASA's blueprint identifies the answers to the question of "Why?" Any single answer is unlikely to satisfy all stakeholders, but each answer is important to one or more stakeholders. Starting with "Why?" will help anchor the development process, but architecture choices may still vary widely depending on how the many different answers to "Why?" are prioritized. Must the first human Mars mission check off *every* item in the "Why?" Venn diagram, or is it sufficient to establish a first-mission architecture that meets the highest-priority items, and is extensible to meet lower priorities during subsequent missions?

For example, prioritizing Science on the first human Mars mission will influence "Where" we land if the specific science objective of interest requires access to a particular region or feature and may require mission elements tailored to that particular science discipline. If that priority science location is difficult to reach or lacks the resources for sustained human presence, a lighter exploration footprint may be desired for the first mission, and crew selection may be heavily influenced by science expertise. Conversely, if Inspiration, in the form of sustained human presence, is the priority goal, then a landing site offering abundant resources or ease of access may be desired, with the first mission elements laying the groundwork for a heavier, permanent infrastructure at a single location, able to support a larger number of crew, possibly selected for their engineering expertise. As shown in Figure 2-5, different priorities within "Why?" will cascade through the other questions.

These sample decision structures illustrate an important point: the Mars architecture will heavily depend on which decisions are prioritized and which are allowed to "float" to enable the highest priorities first. In practice, the Mars architecture decision flow is likely to be iterative rather than linear. To minimize disruption, rework, and cost or schedule changes, understanding the minimum goals and priorities for the first mission, as well as the longer-term goals for subsequent missions, can aid in establishing a flexible and sustainable architecture. The answer to any one of these questions is less important than whether the answers to all six complement one another as a set and can be balanced to establish an architecture that is achievable, affordable, and adaptable.

2.3.2 Unique Considerations for Mars

2.3.2.1 Mars Architecture Frame of Reference

In Mars architecture discussions it is helpful to keep in mind that mission distances traveled will be at a scale far beyond the entirety of human space flight experience to date (Figure 2-6). A single round-trip journey between Earth and Mars will put about 1.8 to 2 billion kilometers on a Mars transportation system's odometer, regardless of departure opportunity or trajectory traveled—that is roughly equivalent to 950 round trips to the Moon. The distance that the Moon circles Earth only varies by about 43 thousand kilometers over time, so it always takes about the same amount of energy to travel to the Moon and back no matter when we go. By contrast, the distance between Earth and Mars can vary by as much as 340 million kilometers as the two planets orbit the Sun. The closest Mars ever approaches Earth is 54.6 million kilometers; at their farthest, over 400 million kilometers of deep space separates the two planets. This means that much of the operational experience and many of the paradigms—such as mission control, sparing/resupply strategy, crew rescue, or mission abort contingency planning—will require a different approach than previously used on heritage programs (such as the ISS).

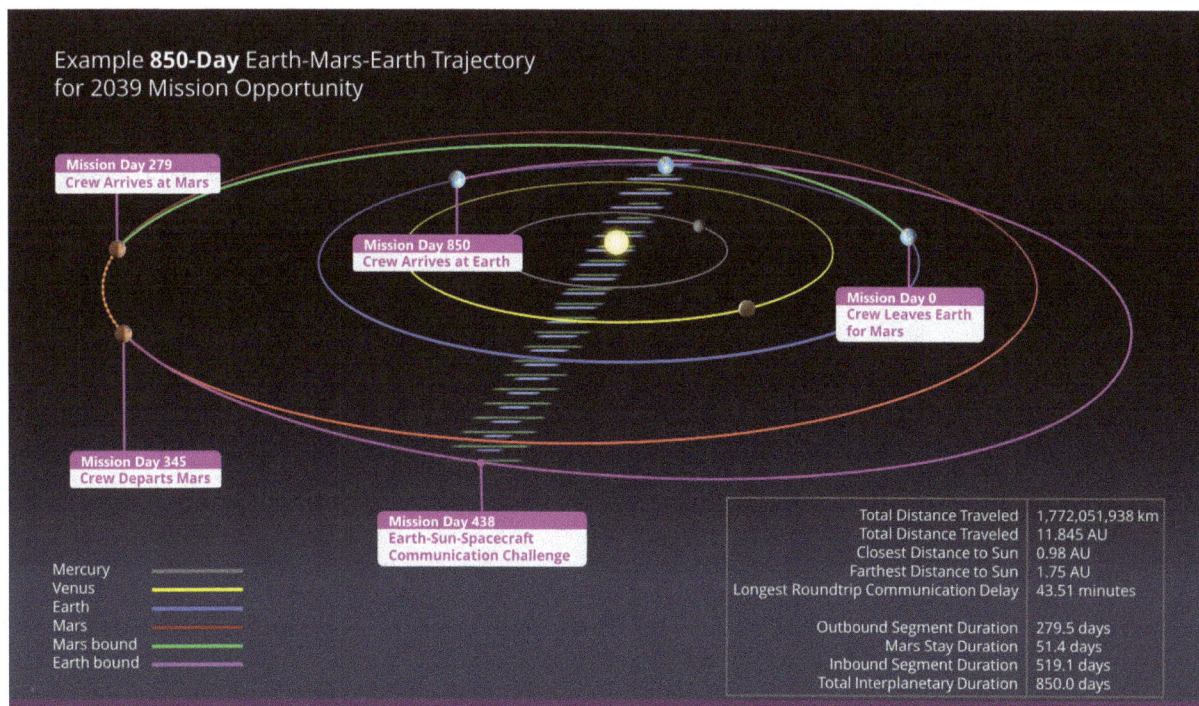

Example **850-Day** Earth-Mars-Earth Trajectory for 2039 Mission Opportunity

Mission Day 279
Crew Arrives at Mars

Mission Day 850
Crew Arrives at Earth

Mission Day 0
Crew Leaves Earth for Mars

Mission Day 345
Crew Departs Mars

Mission Day 438
Earth-Sun-Spacecraft
Communication Challenge

Mercury
Venus
Earth
Mars
Mars bound
Earth bound

Total Distance Traveled	1,772,051,938 km
Total Distance Traveled	11.845 AU
Closest Distance to Sun	0.98 AU
Farthest Distance to Sun	1.75 AU
Longest Roundtrip Communication Delay	43.51 minutes
Outbound Segment Duration	279.5 days
Mars Stay Duration	51.4 days
Inbound Segment Duration	519.1 days
Total Interplanetary Duration	850.0 days

Figure 2-6. Roundtrip Mars Mission Distance In Perspective (AU, Astronomical Unit)

To transit from Earth to Mars and back, the energy required to achieve the roundtrip journey is highly dependent on the timing. Because both planets are orbiting around the Sun, both the distance and the relative velocity of the planets are constantly changing, cycling on a roughly 15- to 20-year cycle. It always takes about the same amount of energy to reach the Moon from Earth, but the amount of energy required to reach Mars varies considerably over this cycle. As part of the "When?" decision, a determination must be made on whether to optimize the transportation system for the easiest opportunities (more affordable but limits us to one mission every 15 to 20 years) or optimize for the most difficult opportunities (less affordable, but allows missions every 2 years), or aim for something in the middle.

Traditionally, to minimize the total energy required to achieve the roundtrip mission, mission planning has selected optimal planetary departure and arrival timing to maximize the benefit of the natural relative position and velocity between the planets. This results in what is typically known as conjunction-class long-stay missions, where both the Earth-to-Mars and Mars-to-Earth trajectories are minimum-energy in nature, typically 180–300 days in duration (each way) depending on the mission opportunity. This approach requires a Mars stay time of 300–500 days to wait for the proper planetary alignment for the return trip and results in a roundtrip total mission duration of around three years.

To achieve shorter duration roundtrip missions to Mars, less-energy-efficient trajectories must be utilized. The energy vs. time trade for a roundtrip mission to Mars is a continuum, but the relationship is exponential in nature: as the mission duration is shortened, the energy required to achieve the roundtrip mission increases exponentially. This translates to an exponential increase in the vehicle mass required, in terms of both propellant and propulsion system, to achieve the roundtrip journey. The total energy required is also highly dependent on the Mars stay time. Unlike the minimum-energy conjunction-class mission, where the Mars stay time is dictated by the waiting period for the optimal return trajectory, shorter roundtrip missions do not have built-in constraints for Mars stay time. This design parameter becomes a key driving factor in interplanetary mission planning. Shorter mission duration also results in shorter stay time at Mars.

An example of these shorter roundtrip missions to Mars is an opposition-class short-stay mission. This class of roundtrip mission to Mars is optimized with one minimum-energy transit (either Earth-to-Mars or Mars-to-Earth), like the conjunction-class missions, and one high-energy transit that is timed to take advantage of a gravity assist swing-by of Venus during opportunities where Venus is in the correct location. This trajectory has typical transit time of 180–300 days each way, with a very short Mars stay time between 10 and 50 days, to achieve a roundtrip total mission duration as short as two years.

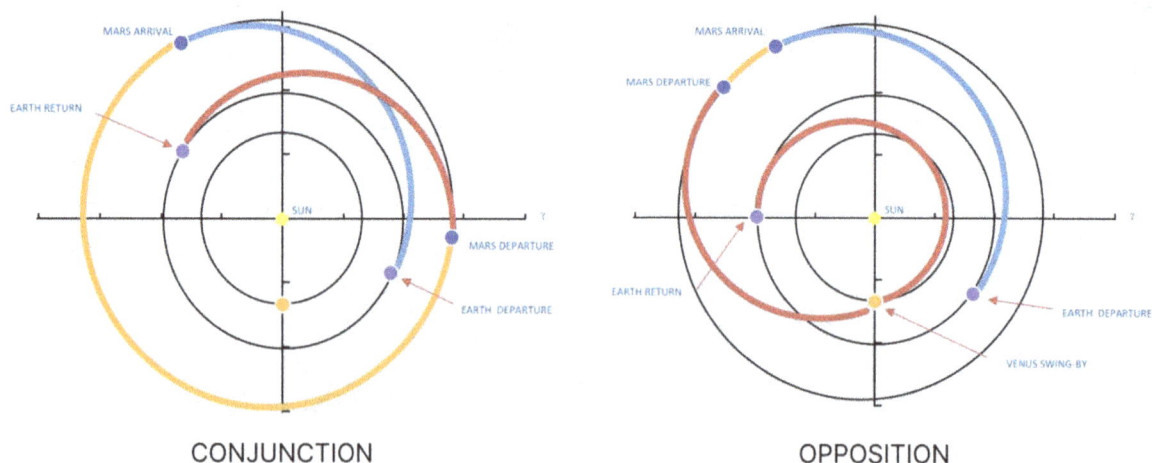

CONJUNCTION   OPPOSITION

Figure 2-7. Illustration of Conjunction- and Opposition-Class Mars Trajectories

These two classes of mission have traditionally been the focus of Mars mission design and planning, but it is important to note that roundtrip missions to Mars are not limited to these two options, as evinced by the example trajectory shown in Figure 2-7. Mars mission design should not be a contest of "conjunction" vs. "opposition," but rather an integrated and thoughtful analysis of all parameters of interest. Roundtrip transit time, Mars stay time, and departure dates are all important factors in determining the total energy required to achieve roundtrip missions. Analyzing the implications of each factor on all relevant systems will help better understand the overall design trade space to support more informed decisions.

2.3.2.2 Aggregate Mars Mission Risk

Throughout the entire 60-year history of human space flight, astronauts have never been more than a few days (and rarely more than a few hours) from Earth. For missions to the International Space Station, or even to the Moon, aborting the mission and returning home is a relatively straightforward option. But on the transit to Mars, mission abort is complicated because the sheer distance between Earth and Mars will take significant time to cover. Depending on when abort is initiated in the mission timeline, the heliocentric nature of the transit may require *months* to return to Earth, regardless of the transportation system selected. For transportation architectures that rely on Mars vicinity return fueling strategies, mission abort during the outbound transit leg may not be possible. In many cases, transit abort will not be a practical response to an emergency because the time to effect crew return will exceed the amount of time within which the crew must resolve the emergency. Early human Mars missions will also have limited Mars ascent/descent abort options. Mars atmosphere and gravity make it difficult to carry sufficient on-board propellant to initiate human-scale payload descent and abort back to orbit during Mars descent, and Mars will initially lack the specialized infrastructure and staffing needed to aid crew after an ascent abort back to the Mars surface—even a successful abort to the surface may very well leave crew stranded away from assets necessary for a safe return to Mars orbit. These challenges will require an entirely new contingency operations paradigm for initial human Mars missions relative to our Earth-centric flight experience. Given that crew survival has been key in meeting human-rating certification loss-of-crew requirements (as derived from Administrator-established safety risk thresholds), additional emphasis will need to be put upon hazard mitigation via other measures (e.g., incorporation of additional reliability and maintainability of hardware/software and a heavier reliance upon autonomy into the architecture) to do the same for a Mars architecture. Such measures will need to account for various other factors, including longer Earth-based communication delays and blackout periods, negative mental health and physiological impacts of transit and surface ops, and impacts upon human reliability.

The farther that humans travel from Earth, the more risk we must accept to achieve the goals of exploration. Mission durations, travel distances, and mass constraints increase the probabilities of something not performing as expected and decrease our ability to respond in a timely manner to emergencies. Crew health, safety, and survival techniques will necessarily change as we move into Mars exploration. The definition of and acceptance of reasonable levels of risk will be a driving factor in determining architecture capabilities and use cases. The definition of acceptable risk is influenced heavily by both internal and external environments and, thus, must be explicitly defined and understood within the architecture so that it can influence decisions throughout the design and implementation process.

2.3.2.3 The Human System in the Mars Architecture

What is often lacking in Mars architecture discussions is a recognition that the human system must be considered as part of the integrated mission architecture. Historically, emphasis on conjunction class Mars missions on the order of 3 or more years duration was driven by a desire to lower Earth-launched transit propellant mass. While this may result in a "better" architecture from a transportation system point of view—with total stack mass serving as the measuring stick for "better"—the 3-or-more-year conjunction-class mission duration is not necessarily better from a crew health and performance perspective. From a purely medical point of view, it would seem intuitively obvious that the 2-year opposition-class mission should be "better" for the crew than the longer-duration conjunction class mission due to the shorter time spent in the deep space environment, but that conclusion is premature without more insight into the integrated vehicle risks that will be layered on top of the medical risks, as well as considerations for crew performance. Beyond the transportation and habitation systems, crew support elements, such as a long-duration food system, remote medical care, laundry/clothing, on-demand training aids, communications, physical and psychological support, and utilization systems, must be included as part of the end-to-end human Mars architecture.

To ensure the human system is well integrated into the overall architecture, NASA is exercising a process to develop more robust spaceflight systems and build a culture of interplanetary human exploration, guided by the Agency's new Blueprint Objectives for exploration. This process incorporates iterative steps building on lessons learned from NASA assets and operations—such as Earth analogs, ISS and commercial LEO missions, and the development of plans for Artemis— to mature plans for future human Mars missions and to use these plans to inform activities for ISS and future platforms in low-Earth orbit, as well as Gateway and lunar surface mission analogs during upcoming Artemis missions. The knowledge gained from these will reduce uncertainty and risk for Mars.

2.3.2.4 Mars Architecture Development Approach

The light-footprint initial mission architecture that has been developed over the past several analysis cycles will serve as a starting point to define one corner of the trade space. This modest architecture concept, described in more detail below, will be expanded through a methodical process to develop the first human Mars mission campaign. The decomposition of the Agency objectives will drive the specific functions and use cases that will inform Mars architecture strategy. NASA will coordinate with stakeholders to explore integrated architecture impacts, such as how infrastructure or science objectives influence mass, volume, power, and overall transportation and habitation design. This will be an iterative process, resulting in a catalog of key Mars architecture decisions. To minimize disruption, rework and cost or schedule changes, the Mars architecture will be assessed from each of the six key questions ("Who," "What," "Where," etc.) noted above to understand how different decision structures influence the architecture before selecting a development starting point that best fits the Agency's M2M Strategy. Where additional research is required to inform a decision, NASA will coordinate activities across the Agency, which may include testing, analysis, or analog investigations on Earth, orbiting platforms, or the lunar surface. Objectives will be prioritized to align with anticipated resource availability timelines, opportunities, and partner agreements. As a roadmap of key architecture decisions emerges and the trade space is narrowed, this document will be updated to reflect the evolving Mars architecture.

Since the Mars architecture will be built up over time, interoperability is a vital aspect to ensure compatibility between elements and systems. With limited ground support, on-board autonomy (crew and systems) as well as interoperability in the Mars campaign will be crucial for crew safety and mission success. Lunar interoperability lessons will guide the development of interoperability for Mars architecture systems. Compatible systems envisioned include deep space vehicles, surface vehicles, utilization/science, and logistics operations. Mars architecture will work with lunar programs to evaluate best practices learned from the lunar campaign as well as define the future needs for specific system compatibility.

DECOMPOSITION OF OBJECTIVES

M Strategy[6] document contains the overarching goals and objectives of the long-term development leading to exploratio n, Mars, and beyond. As described in Section 1.3.1, the decomposition of the M2M Strategy begins with the definition c ristics and needs that translates the objectives into high-level actionable exploration activities in support of fu ositions into the specific use cases and functions. The Agency's top-level objective owners are key contributors of the defir characteristics and needs as applicable to the human exploration architecture. The characteristics and needs are defin vel with the intent of not specifying explicit solutions to produce the desired results. The objectives are also accompani I Strategy by nine Recurring Tenets that reflect common themes across the objectives. These themes present principle in which the objective satisfaction will occur. These Recurring Tenets are replicated in Section 4.0, in addition tc nents of the M2M architecture adherence to them.

Lunar Objective Decomposition to Characteristics and Needs

Science

is broken down into six individual areas, each with its own goal statement: Lunar/Planetary Science, Heliophysics Scie and Biological Science, Physics and Physical Science, Science-Enabling, and Applied Science

Lunar/Planetary Science (LPS)

ddress high priority planetary science questions that are best accomplished by on-site human explorers on and aroun d Mars, aided by surface and orbiting robotic systems.

Moon-to-Mars Strategy and Objectives Development. NP-2023-03-3115-HQ.

...racteristics & Needs	←	Objective
...sit geographically diverse sites around the South Pole and non-polar regions ...entify and collect samples from multiple locations, including frozen samples from PSRs, the lunar surface ...turn a variety of types of samples collected from the lunar surface, including of soil, ...bbles, and rocks (hand sample size) ...turn a variety of samples from the lunar subsurface ...turn select samples of regolith, rock, and/or subsurface materials in containers sealed ...unar vacuum ...nplace and operate science packages at distributed sites on the lunar surface ...ility to know where samples are collected from ...ploy scientific payloads at distances outside the blast zone of the ascent vehicle ...ovide power to deployed science payloads to enable sustained operation for durations ...several years ...ility for the Science Evaluation Room to exchange data and interact with crew in real ...e	←	Uncover the record of solar system origin and early history, by determining how and when planetary bodies formed and differentiated, characterizing the impact chronology of the inner solar system as recorded on the Moon and Mars, and characterize how impact rates in the inner solar system have changed over time as recorded on the Moon and Mars.
...sit geographically diverse sites around the South Pole and non-polar regions ...entify and collect samples from multiple locations, including frozen samples from PSRs, the lunar surface ...turn a variety of types of samples collected from the lunar surface, including of soil, ...bbles, and rocks (hand sample size) ...turn a variety of samples from the lunar subsurface ...turn select samples of regolith, rock, and/or subsurface materials in containers sealed ...unar vacuum ...nplace and operate science packages at distributed sites on the lunar surface ...ility to know where samples are collected from ...ploy scientific payloads at distances outside the blast zone of the ascent vehicle ...ovide power to deployed science payloads to enable sustained operation for durations ...several years ...ility for the Science Evaluation Room to exchange data and interact with crew in real ...e	←	Advance understanding of the geologic processes that affect planetary bodies by determining the interior structures, characterizing the magmatic histories, characterizing ancient, modern, and evolution of atmospheres/exospheres, and investigating how active processes modify the surfaces of the Moon and Mars.

teristics & Needs	←	Objective	ID
eographically diverse sites around the South Pole and non-polar regions y and collect samples from multiple locations, including frozen samples from PSRs, lunar surface a variety of types of samples collected from the lunar surface, including of soil, s, and rocks (hand sample size) a variety of samples from the lunar subsurface select samples of regolith, rock, and/or subsurface materials in containers sealed r vacuum ce and operate science packages at distributed sites on the lunar surface to know where samples are collected from scientific payloads at distances outside the blast zone of the ascent vehicle e power to deployed science payloads to enable sustained operation for durations eral years for the Science Evaluation Room to exchange data and interact with crew in real	←	Reveal inner solar system volatile origin and delivery processes by determining the age, origin, distribution, abundance, composition, transport, and sequestration of lunar and Martian volatiles.	LI

2 Heliophysics Science (HS)

ddress high priority Heliophysics science and space weather questions that are best accomplished using a combinati explorers and robotic systems at the Moon, at Mars, and in deep space.

teristics & Needs	←	Objective	ID
ce and operate science instrumentation in a variety of lunar orbits ce and operate science instrumentation for solar monitoring off the Earth-Sun line ce and operate science instrumentation on the lunar surface e power, communications, and data to deployed science payloads to enable ned operation for durations of several years	←	Improve understanding of space weather phenomena to enable enhanced observation and prediction of the dynamic environment from space to the surface at the Moon and Mars.	H
nent and collect samples from multiple locations on the lunar surface to Earth drill core samples of lunar soil collected from the lunar subsurface at a of depths ce and operate science instrumentation on the lunar surface	←	Determine the history of the Sun and solar system as recorded in the lunar and Martian regolith.	H
ce and operate science instrumentation in a variety of lunar orbits ce and operate science instrumentation in globally distributed locations on the lunar e nent and collect samples from globally distributed locations on the lunar surface	←	Investigate and characterize fundamental plasma processes, including dust-plasma interactions, using the cislunar, near-Mars, and surface environments as laboratories.	H

racteristics & Needs	←	Objective
place and operate science packages in lunar orbit place and operate science instrumentation on the lunar surface	←	Improve understanding of magnetotail and pristine solar wind dynamics in the vicinity of the Moon and around Mars.

.1.3 Human and Biological Science (HBS)

: Advance understanding of how biology responds to the environments of the Moon, Mars, and deep space to amental knowledge, support safe, productive human space missions and reduce risks for future exploration.

racteristics & Needs	←	Objective
vide numerous mid-duration crew increments on the lunar surface vide numerous long duration crew increments in cis lunar space prior to surface ssion nduct crew transitions from micro-gravity to partial gravity	←	Understand the effects of short- and long-duration exposure to the environments of the Moon, Mars, and deep space on biological systems and health, using humans, model organisms, systems of human physiology, and plants.
vide numerous mid-duration crew increments on the lunar surface vide numerous long duration crew increments in cis lunar space prior to surface ssion nduct crew transitions from micro-gravity to partial gravity	←	Evaluate and validate progressively Earth-independent crew health & performance systems and operations with mission durations representative of Mars-class missions.
vide numerous mid-duration crew increments on the lunar surface vide numerous long duration crew increments in cis lunar space prior to surface ssion nduct crew transitions from micro-gravity to partial gravity	←	Characterize and evaluate how the interaction of exploration systems and the deep space environment affect human health, performance, and space human factors to inform future exploration-class missions.

.1.4 Physics and Physical Science (PPS)

: Address high priority physics and physical science questions that are best accomplished by using unique attributes of t onment.

racteristics & Needs	←	Objective
place and operate astrophysics science instrumentation on the far side lunar surface \ facilities (e.g., instruments, racks, stowage, power) on the lunar surface to enable damental physics experiments	←	Conduct astrophysics and fundamental physics investigations of deep space and deep time from the radio quiet environment of the lunar far side.
place and operate science packages in lunar orbit place and operate science packages on the lunar surface vide IVA laboratory space on the lunar surface and provide for crew time to conduct eriments	←	Advance understanding of physical systems and fundamental physics by utilizing the unique environments of the Moon, Mars, and deep space.

Science-Enabling (SE)

integrated human and robotic methods and advanced techniques that enable high-priority scientific questions to be addre
and on the Moon and Mars.

eristics & Needs	←	Objective	ID
stronauts to be field geologists and to perform additional science tasks during s missions, through field and classroom training	←	Provide in-depth, mission-specific science training for astronauts to enable crew to perform high-priority or transformational science on the surface of the Moon, and Mars, and in deep space.	SE
ent communications systems to enable high bandwidth, high availability inications between Earth-based personnel, surface crew, and science packages on face Earth-based scientists, integrated with FOD, to support crew activities in real time	←	Enable Earth-based scientists to remotely support astronaut surface and deep space activities using advanced techniques and tools.	SE
to Earth frozen samples from a variety of depths in their pristine state to JSC n facilities select samples of soil, rock, and/or subsurface materials in containers sealed in acuum to JSC curation facilities y tools, including temperature sensors, to support acquisition of frozen samples, ctured in accordance with science requirements to minimize sample contamination y sample containers appropriate for the specimens collected and science needs ontamination considerations), including sealed containers and drill core tubes	←	Develop the capability to retrieve core samples of frozen volatiles from permanently shadowed regions on the Moon and volatile-bearing sites on Mars and to deliver them in pristine states to modern curation facilities on Earth.	SE
eologically diverse sites around the South Pole and non-polar regions ent and collect samples from multiple locations, including frozen samples from on the lunar surface to Earth a variety of types of samples collected from the lunar surface, including of ebbles, and rocks (hand sample size) to Earth a variety of samples from the lunar subsurface at a variety of depths y tools to support acquisition of samples, including soil, pebbles, hand-sized rock es, and drill cores, manufactured in accordance with science requirements to ze sample contamination y sample containers appropriate for the specimens collected and science needs ontamination considerations), including bags, sealed containers, and drill core tubes	←	Return representative samples from multiple locations across the surface of the Moon and Mars, with sample mass commensurate with mission-specific science priorities.	SE
c surveys of potential landing sites, including video and in situ measurements y robotic tools to support acquisition of samples, including soil, pebbles, hand-sized mples, and drill cores, manufactured in accordance with science requirements to ze sample contamination y sample containers appropriate for the specimens collected and science needs ontamination considerations), including bags, and sealed containers, that are ible to robotic manipulation	←	Use robotic techniques to survey sites, conduct in-situ measurements, and identify/stockpile samples in advance of and concurrent with astronaut arrival, to optimize astronaut time on the lunar and Martian surface and maximize science return.	SE

racteristics & Needs	←	Objective
place and operate science instrumentation in lunar and heliocentric orbits relevant to dressing the associated science objectives place and operate science instrumentation on the lunar surface at locations relevant to dressing associated science objectives, including polar and non-polar locations on the ar near side and far side	←	Enable long-term, planet-wide research by delivering science instruments to multiple science-relevant orbits and surface locations at the Moon and Mars.
eserve radio free environment on far side nit contamination of PSRs otect sites of historic significance	←	Preserve and protect representative features of special interest, including lunar permanently shadowed regions and the radio quiet far side as well as Martian recurring slope lineae, to enable future high-priority science investigations.

.1.6 Applied Science (AS)

: Conduct science on the Moon, in cislunar space, and around and on Mars using integrated human and robotic meth
nced techniques, to inform design and development of exploration systems and enable safe operations.

racteristics & Needs	←	Objective
place and operate science instrumentation in lunar and heliocentric orbits relevant to dressing the associated science objectives place and operate science instrumentation on the lunar surface at locations relevant to dressing associated science objectives, including polar and non-polar locations on the ar near side and far side	←	Characterize and monitor the contemporary environments of the lunar and Martian surfaces and orbits, including investigations of micrometeorite flux, atmospheric weather, space weather, space weathering, and dust, to plan, support, and monitor safety of crewed operations in these locations.
place and operate science instrumentation in lunar and heliocentric orbits relevant to dressing the associated science objectives place and operate science instrumentation on the lunar surface at locations relevant to dressing associated science objectives, including polar and non-polar locations on the ar near side and far side	←	Coordinate on-going and future science measurements from orbital and surface platforms to optimize human-led science campaigns on the Moon and Mars.

...teristics & Needs	←	Objective	ID
...eologically diverse sites around the South Pole ...x samples from multiple locations, including frozen samples from PSRs, on the lunar ...e ... to Earth a variety of types of samples collected from the lunar surface, including of ...ebbles, and rocks ... to Earth a variety of samples from the lunar subsurface at a variety of depths ... to Earth frozen samples from a variety of depths in their pristine state ... select samples of soil, rock, and/or subsurface materials in containers sealed in ...acuum ...y to conduct prospecting traverses with appropriate scientific instrumentation and ...pabilities over sites of interest ... scientific payloads at distances outside the blast zone of the ascent vehicle ...ce and operate science instrumentation in lunar and heliocentric orbits relevant to ...sing the associated science objectives	←	Characterize accessible lunar and Martian resources, gather scientific research data, and analyze potential reserves to satisfy science and technology objectives and enable In-Situ Resource Utilization (ISRU) on successive missions.	A...
...nstrate operation of bioregenerative ECLSS sub-systems in deep space	←	Conduct applied scientific investigations essential for the development of bioregenerative-based, ecological life support systems	A...
...nstrate operation of plant based ECLSS sub-systems in deep space.	←	Define crop plant species, including methods for their productive growth, capable of providing sustainable and nutritious food sources for lunar, Deep Space transit, and Mars habitation.	A...
...ce and operate science packages in lunar orbit ...ce and operate science packages on the lunar surface ...e IVA laboratory space on the lunar surface and provide for crew time to conduct ...ments	←	Advance understanding of how physical systems and fundamental physical phenomena are affected by partial gravity, microgravity, and general environment of the Moon, Mars, and deep space transit.	A...

.2   Lunar Infrastructure (LI)

: Create an interoperable global lunar utilization infrastructure where U.S. industry and international partners can
nuous robotic and human presence on the lunar surface for a robust lunar economy without NASA as the sole use
mplishing science objectives and testing for Mars.

racteristics & Needs	←	Objective
nplace power generation and power storage capabilities on the lunar surface nplace power distribution and storage capabilities on the lunar surface to allow power ization at multiple locations around exploration sites	←	Develop an incremental lunar power generation and distribution system that is evolvable to support continuous robotic/human operation and is capable of scaling to global power utilization and industrial power levels.
plement communications systems to enable high bandwidth, high availability mmunications between Earth-based personnel, surface crew, and science packages on surface	←	Develop a lunar surface, orbital, and Moon-to-Earth communications architecture capable of scaling to support long term science, exploration, and industrial needs.
plement navigation and timing systems to enable high availability navigation on the face	←	Develop a lunar position, navigation and timing architecture capable of scaling to support long term science, exploration, and industrial needs.
liver and demonstrate autonomous construction demonstration package(s) to the lunar uth Pole	←	Demonstrate advanced manufacturing and autonomous construction capabilities in support of continuous human lunar presence and a robust lunar economy.
monstrate ability of lunar landers to reliably land within a defined radius around an ended location.	←	Demonstrate precision landing capabilities in support of continuous human lunar presence and a robust lunar economy.
monstrate the ability to allow crew to move locally around landing sites to visit multiple ations of interest monstrate the ability to relocate surface elements to locations around the lunar South le between crewed surface missions.	←	Demonstrate local, regional, and global surface transportation and mobility capabilities in support of continuous human lunar presence and a robust lunar economy.
liver and demonstrate ISRU demonstration package(s) to the lunar South Pole	←	Demonstrate industrial scale ISRU capabilities in support of continuous human lunar presence and a robust lunar economy.
monstrate capability to transfer propellant from one spacecraft to another in space monstrate capability to store propellant for extended durations in space monstrate capability to store propellant on the lunar surface for extended durations	←	Demonstrate technologies supporting cislunar orbital/surface depots, construction and manufacturing maximizing the use of in-situ resources, and support systems needed for continuous human/robotic presence.

eristics & Needs	←	Objective	ID
ce systems to monitor solar weather and to predict SPEs	←	Develop environmental monitoring, situational awareness, and early warning capabilities to support a resilient, continuous human/robotic lunar presence.	LI

Transportation & Habitation

evelop and demonstrate an integrated system of systems to conduct a campaign of human exploration missions to the N
s, while living and working on the lunar and Martian surface, with safe return to Earth.

eristics & Needs	←	Objective	ID
strate transportation of crew and systems from Earth to stable lunar orbit istrate staged operation of crew transportation from stable lunar orbit with ibility to both Earth and the lunar South Pole istrate crew transport from stable lunar orbit to lunar surface and from lunar surface IO e crew transportation system in uncrewed mode for extended periods on the lunar istrate safe return to Earth of crew and systems from stable lunar orbit	←	Develop cislunar systems that crew can routinely operate to and from lunar orbit and the lunar surface for extended durations.	TH
strate capabilities to deliver elements from Earth to the lunar surface istrate unloading of cargo from delivery system(s)	←	Develop system(s) that can routinely deliver a range of elements to the lunar surface.	TH
istrate capabilities to allow crew to live on the surface of the Moon istrate capabilities to allow crew to exit habitable space and conduct EVA activities istrate capabilities to allow crew to conduct science and utilization activities	←	Develop system(s) to allow crew to explore, operate, and live on the lunar surface and in lunar orbit with scalability to continuous presence; conducting scientific and industrial utilization as well as Mars analog activities.	TH
istrate in-space assembly of spacecraft in a stable lunar orbit with minimal orbital nance and support for power generation istrate capabilities to allow crew to live in deep space for extended durations	←	Develop in-space and surface habitation system(s) for crew to live in deep space for extended durations, enabling future missions to Mars.	TH
istrate remote crew health sub-system(s) in NRHO istrate remote crew health sub-system(s) on the lunar surface	←	Develop systems that monitor and maintain crew health and performance throughout all mission phases, including during communication delays to Earth, and in an environment that does not allow emergency evacuation or terrestrial medical assistance.	TH
ct operations in which robotic sub-systems, controlled remotely from Earth, NRHO, surface, support crew exploration	←	Develop integrated human and robotic systems with inter-relationships that enable maximum science and exploration during lunar missions.	TH

racteristics & Needs	←	Objective
monstrate capabilities to return cargo from the lunar surface back to Earth	←	Develop systems capable of returning a range of cargo mass from the lunar surface to Earth, including the capabilities necessary to meet scientific and utilization objectives.

1.4   Operations

: Conduct human missions on the surface and around the Moon followed by missions to Mars. Using a gradual build-up a̶ missions will demonstrate technologies and operations to live and work on a planetary surface other than Earth, wit̶ n to Earth at the completion of the missions.

racteristics & Needs	←	Objective
nduct mid-duration crew exploration on the lunar surface nduct long-duration crew exploration in cis lunar space nduct crewed and uncrewed testing of in-space habitation systems	←	Conduct human research and technology demonstrations on the surface of Earth, low Earth orbit platforms, cislunar platforms, and on the surface of the moon, to evaluate the effects of extended mission durations on the performance of crew and systems, reduce risk, and shorten the timeframe for system testing and readiness prior to the initial human Mars exploration campaign.
ovide onboard autonomy to train, plan, and execute a safe mission ovide flight control and mission integration to ensure safety and mission success	←	Optimize operations, training and interaction between the team on Earth, crew members on orbit, and a Martian surface team considering communication delays, autonomy level, and time required for an early return to the Earth.
sit geographically diverse sites around the South Pole and non-polar regions nplace science packages at distributed sites on the lunar surface ntify and collect samples from multiple locations, including frozen samples from PSRs, the lunar surface turn samples collected on the surface, including frozen samples, to Earth	←	Characterize accessible resources, gather scientific research data, and analyze potential reserves to satisfy science and technology objectives and enable use of resources on successive missions.
egrate networks and mission systems to exchange data between Earth and campaign stems	←	Establish command control processes, common interfaces, and ground systems that will support expanding human missions at the Moon and Mars.
ansport crew and cargo between landing or base site and exploration sites of varying stances nduct EVA activities utilizing mobility assets and tools	←	Operate surface mobility systems, e.g., extra-vehicular activity (EVA) suits, tools and vehicles.
nduct mid-duration crew exploration on the lunar surface nduct long-duration crew exploration in cis lunar space ansition crew from micro-gravity to partial gravity	←	Evaluate, understand, and mitigate the impacts on crew health and performance of a long deep space orbital mission, followed by partial gravity surface operations on the Moon.

teristics & Needs	←	Objective	ID
ct extended crewed and uncrewed testing of in-space habitation system e crew health and performance capabilities for Mars duration mission e crew survival capabilities	←	Validate readiness of systems and operations to support crew health and performance for the initial human Mars exploration campaign.	O
s and re-use surface assets from previous crewed and uncrewed missions	←	Demonstrate the capability to find, service, upgrade, or utilize instruments and equipment from robotic landers or previous human missions on the surface of the Moon and Mars.	O
e autonomous and remote operations of surface systems from external systems, ng Earth, orbital, and other surface locations e safe interactions between crew and automated/autonomous systems	←	Demonstrate the capability of integrated robotic systems to support and maximize the useful work performed by crewmembers on the surface, and in orbit.	O
e autonomous and remote operations of surface systems from external systems, ng Earth, orbital, and other surface locations e safe interactions between crew and automated/autonomous systems	←	Demonstrate the capability to operate robotic systems that are used to support crew members on the lunar or Martian surface, autonomously or remotely from the Earth or from orbiting platforms.	O
ce and operate ISRU demonstration packages on the lunar surface	←	Demonstrate the capability to use commodities produced from planetary surface or in-space resources to reduce the mass required to be transported from Earth.	O
ve radio free environment on far side ontamination of PSRs nstrate recovery of excess fluids and gases from lunar landers	←	Establish procedures and systems that will minimize the disturbance to the local environment, maximize the resources available to future explorers, and allow for reuse/recycling of material transported from Earth (and from the lunar surface in the case of Mars) to be used during exploration.	O

Mars Objective Decomposition to Characteristics and Needs

composition of the Mars objectives into the architecture characteristics and needs are yet to be completed. It will follow rocess outlined in this document in future iterations of the ADD. It is understood that the lunar objective decomposi eristics and needs, and definition of Use Cases and Functions will all be influenced by the Mars objectives.

Example Decompositions to Use Cases and Function

ection, an example of the full decomposition from the objective to the Characteristics and Needs to the Use Cases and Func . For the comprehensive decomposition of every lunar objective into the Use Cases and Functions, please refer to APPE Decomposition of Lunar Objectives.

Use Cases	Functions	Characteristics & Needs	Objectives & Go
ransport crew and systems rom Earth to cislunar space	• Provide ground services • Stack and integrate • Manage consumables and propellant • Enable vehicle launch • Provide multiple launch attempts • Provide aborts • Transport crew and systems from Earth to cislunar space	Demonstrate transportation of crew and systems from Earth to stable lunar orbit	**TH-1**
Staging of crewed Lunar rface missions from cislunar space	• Vehicle rendezvous, prox ops, docking, and undocking in cislunar space • Provide PNT capability in cislunar space • Provide crew habitation in cislunar space	Demonstrate staged operation of crew transportation from stable lunar orbit with accessibility to both Earth and the lunar South Pole.	Develop cislunar syster crew can routinely ope and from lunar orbit a lunar surface for exte durations.
Physical assembly of ntegrated assets in cislunar space	• Transport elements from Earth to cislunar space • Docking/berthing of spacecraft elements	Demonstration of crew transport from stable lunar orbit to lunar surface and from lunar surface to NRHO.	
Crew transport between cislunar space and Lunar surface	• Transport crew and systems from cislunar space to Lunar surface South Pole sites • Transport crew and systems from Lunar surface to cislunar space	Operate crew transportation system in uncrewed mode for extended periods on the lunar surface	
Human lander operates in andby mode while crew live in surface systems	• Operate crew vehicle in uncrewed mode on surface	Demonstrate safe return to Earth of crew and systems from stable Lunar orbit	
Return crew and systems om cislunar space to Earth	• Transport crew and systems from cislunar space to Earth • Recover crew, systems, and cargo after splashdown		

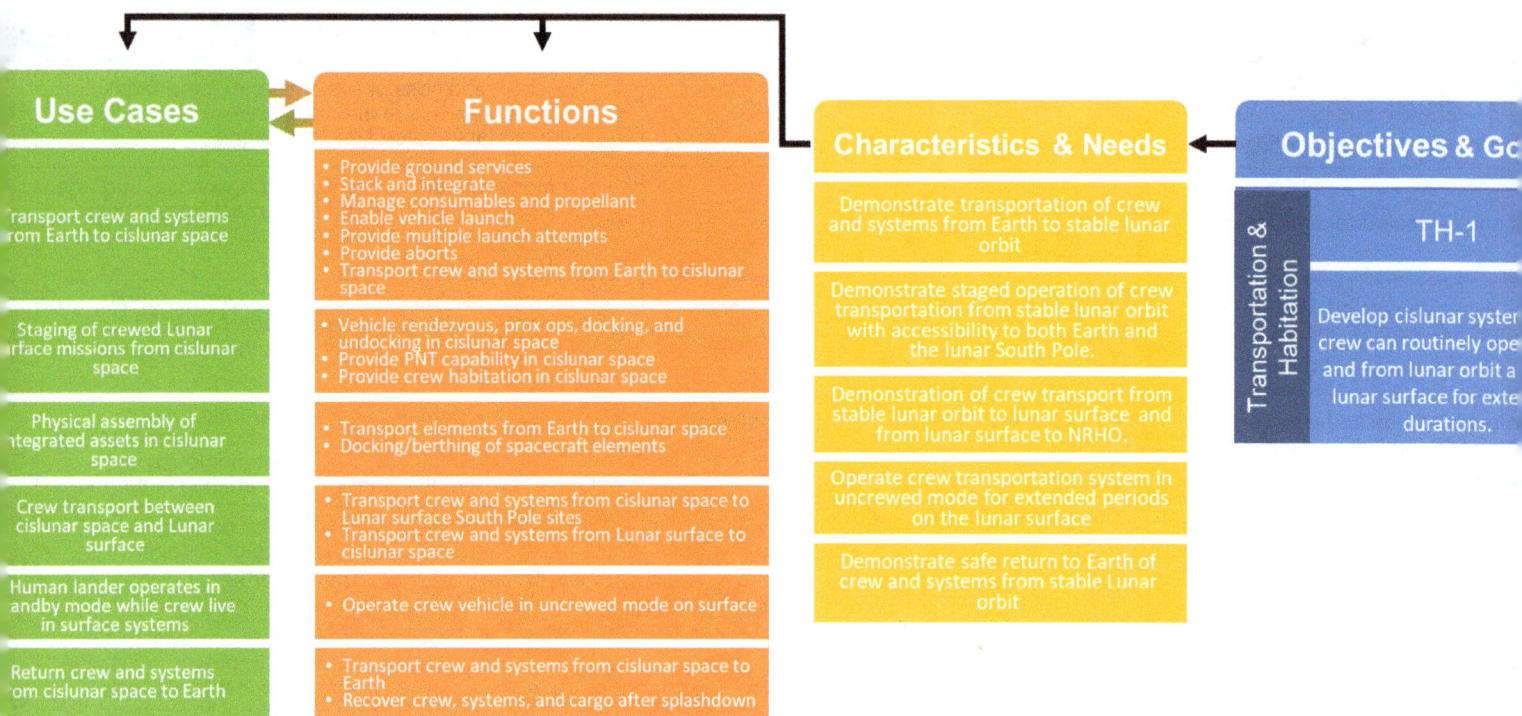

Figure 2-8. Example of the Full Decomposition of the Objectives into Lunar Specific Uses Cases and Functions

e 2-8 shows a complete example, from objective through to Use Cases, for one Objective under the Transportation and Ha TH-1. The overall Objective breaks down into five distinct Characteristics and Needs that are necessary to satisfy ctive, as defined by the Goal owners. Five Use Cases and associated sets of Functions are then identified that would Characteristics and Needs.

3.0   MOON-TO-MARS ARCHITECTURE

The architecture methodology process described in Section 1.3 has yielded a structured approach to objective decomposition and applicability to system definition to establish the architecture.

Return—The architecture starts with the development and demonstration of the systems that transport crew and exploration capabilities to target destinations. The successful Artemis I mission was the first step in this progressive expansion of the capability envelope over a series of missions where a minimum crew of 4 can support missions in deep space and on the lunar surface, and eventually future destinations.

Explore—Using an evolutionary approach, the architecture enables high priority science, technology demonstrations, systems validation, and operations for crew to live and work on a non-terrestrial planetary surface, with a safe return to Earth at the completion of the mission(s). Key characteristics include operating and designing the lunar systems with Mars risk reduction in mind, from a systems, operations, and human perspective. The architecture accommodates this approach in the context of available capabilities and differences in the lunar and Mars environments. Initially this is done at the element level, then through combined operations that eventually culminate in several precursor missions in the lunar vicinity where the crew experiences long durations in the deep space environment coupled with rapid acclimation to partial gravity excursions using Mars-like systems and operations. The Mars-forward exploration systems also have the goal to maximize crew efficiency for utilization, which will be tested by a continuum of excursions to a diverse set of sites driven by science needs. The balance between diverse site access and long-duration infrastructure objectives will inform the allocation of functions across systems.

Sustain—The foundational exploration capabilities serve as a basis to increase global access, industrial-scale ISRU, and crew durations beyond NASA's initial needs. Although evolution of the lunar architecture along the lines of these greater capabilities would seem to occur later in the architecture, the implications of the potential future lunar states are initiated at the very beginning of the architecture to the early reconnaissance missions, where something like volatiles access and purity in several regions may dictate the role and level of ISRU.

The lunar architecture is developing, deploying, and operating systems for lunar vicinity exploration; performing science at diverse locations, returning lunar samples; preparing for further exploration with Mars-capable systems, operations, and precursor missions; and establishing a permanent lunar presence that could one day support a lunar economy. The Mars architecture can follow the same basic approach as the Moon, to achieve a human presence, explore and then sustain development.

The architecture has been structured to reflect the incremental buildup of capabilities and objective satisfaction. These campaign segments have been crafted along the Return, Explore, and Sustain approach to further delineate the continuum of evolving capability and objective satisfaction. They are described in Table 3-1 below. Although the segments appear sequential in the table, they are not exclusively serial, as the segments build upon each other and focus on how systems will work together to achieve objective satisfaction.

Table 3-1. Moon-to-Mars Campaign Segments

Human Lunar Return	Initial capabilities, systems, and operations necessary to re-establish human presence and initial utilization (science, etc.) on and around the Moon.
Foundational Exploration	Expansion of lunar capabilities, systems, and operations supporting complex orbital and surface missions to conduct utilization (science, etc.) and Mars forward precursor missions.
Sustained Lunar Evolution	Enabling capabilities, systems, and operations to support regional and global utilization (science, etc.), economic opportunity, and a steady cadence of human presence on and around the Moon.
Humans to Mars	Initial capabilities, systems, and operations necessary to establish human presence and initial utilization (science, etc.) on Mars and continued exploration.

The initial segment is **Human Lunar Return**. This segment includes the initial capabilities, systems, and operations necessary to re-establish human presence and initial utilization (science, etc.) on and around the Moon. This segment's primary focus is establishing the missions and supporting infrastructure to perform sortie crewed missions to the Moon. The systems and support span Earth, cislunar orbiting platforms, and the foothold capabilities on the lunar surface. The initial support of utilization focuses on the human-conducted science, sample collection, human research, and initial capabilities, among others, for the first time outside low-Earth orbit in over 50 years.

The **Foundational Exploration** segment includes lunar excursions to diverse sites of interest with increasingly higher-complexity missions, enabling science and other utilization exploration. This segment also contributes to evaluating the systems, operations, human adaptation, or technologies required for Mars. These missions will enable increasingly extended time in deep space coupled with missions to the lunar surface of increasing duration and mobility that address identified research, testing, and demonstration objectives to enable Mars missions. Prior to the crewed Mars mission, these precursor missions would be performed in time to inform element design, testing, and operation. Foundational Exploration also starts the development of a sustainable human presence with the deployment of long-term infrastructure.

The third segment, **Sustained Lunar Evolution**, is the broad and undefined end state that builds on the foundation of the first two segments and enables capabilities, systems, and operations to support regional and global utilization (science, etc.), economic opportunity, and a steady cadence of human presence on and around the Moon. Here we can envision various uses of the lunar surface and cislunar space to enable science, commerce, and further deep space exploration initiatives.

The fourth segment, **Humans to Mars**, captures the capabilities, systems, and operations necessary to enable the initial human exploration of the Red Planet. These systems will represent the transportation, logistics, utilization, and more required to enable the missions. This segment is an enabling capability of continued deep space exploration with additional efforts to be identified as architectural progress occurs.

3.1 HUMAN LUNAR RETURN SEGMENT

The Human Lunar Return (HLR) segment of the exploration campaign includes the inaugural Artemis missions to enable returning humans to the Moon and demonstrating both crewed and uncrewed lunar systems, including the support to initial utilization (science, etc.) capabilities. This segment will be used to demonstrate initial systems to validate system performance and to establish a core capability for follow-on campaign segments. It captures the missions that test NASA's deep space crew and cargo transportation system(s), deploy the initial cislunar capabilities to support lunar missions, deploy and establish lunar orbital communication relays, and bring two crew members to the lunar surface, followed by their safe return to Earth. Additionally, a variety of other efforts are in work to support data-gathering and risk-reduction activities to help inform future decisions. These currently include, but are not limited to, the Cislunar Autonomous Positioning System Technology Operations and Navigation Experiment (CAPSTONE), Commercial Lunar Payload Services (CLPS), and Volatiles Investigating Polar Exploration Rover (VIPER).

3.1.1 Summary of Objectives

The objectives that drive the HLR segment include achieving science, inspiration, and national posture goals around and on the surface of the Moon. Initial missions will be used to deliver science value through operations in cislunar space and on the lunar surface, along with the return of samples to Earth. Key science objectives addressable during HLR include 1) exploring the lunar south polar region to understand chronology, composition, and structure of this region (e.g., LPS-1 and LPS-2); 2) understanding volatile composition and the environment of shallow permanently shadowed regions near the lunar south pole (e.g., LPS-3); 3) assessing the history of the Sun as preserved in lunar regolith (e.g., HS-2); 4) characterizing space weather dynamics to enable future forecasting capabilities (e.g., HS-1); and 5) characterizing plant, model organism/systems, and human physiological responses in partial-gravity environments (e.g., HBS-1). These HLR science priorities were identified by the Science Mission Directorate.

In order to achieve these key science, inspiration, and national posture goals, the HLR segment is focused on demonstrating initial capabilities, systems, and operations necessary to re-establish human presence around and on the Moon. This segment began successfully with the Artemis I mission to systematically and progressively test areas such as crewed transportation to cislunar space (TH-1, TH-2), supporting ground infrastructure (OP-4), and deep-space communications and tracking systems (OP-2). The next steps are crewed transportation to and from cislunar space, initial Gateway deployment (OP-6), rendezvous and docking, uncrewed Human Landing System demonstration, initial human landing (TH-2), and initial surface EVA capability, and uncrewed payload delivery. It encompasses the return of humans to the Moon for ~6-day surface missions and establishes the foundational capabilities that will enable future campaign segments.

The objectives linked to the HLR segment will be a subset of the total, and even of those linked, some will be only partially satisfied; however, the segment serves as the starting point to define and validate capabilities and functions in later segments that will be driven by the objectives. The complete set of HLR objectives can be found in Appendix A and denoted by purple text.

3.1.2 Use Cases and Functions

The objectives and mapping to the use cases and functions (shown in Appendix A) are used to drive the elements for this segment. For HLR, as many elements are operational or in design/development stages, these elements form the basis of satisfying the functional needs. The mappings help identify functional gaps that must be addressed in the follow-on segments. Table

3-2 and Table 3-3 show the mapping of the use cases and functions, respectively, to the elements. The "X" in the tables indicate mapping of use cases to elements. The "X" does not indicate that an element fully satisfies the use case, function, or objectives, or that completion is achieved. Many of the use cases and functions will require additional elements or new functional capabilities that go beyond what is being assigned to the HLR elements described below. Key gaps between planned HLR capabilities and M2M objectives needs are noted later in this document and will continue to be expanded through the ACR process. Note that not all use cases (UC-#) and functions (F-#) are sequential in this segment mapping. The numbering represents use cases and functions that have been identified through the overall objective decompositions process but not all are applicable to the HLR segment.

The mapped elements in HLR segment and their corresponding description section are:

- Space Launch System (SLS) — Section 3.1.4.5.1
- Exploration Ground Systems (EGS) — Section 3.1.4.7.1
- Orion — Section 3.1.4.5.2
- Gateway — Section 3.1.4.2.1
- Human Landing System (HLS) — Section 3.1.4.5.3
- Lunar Communications Relay and Navigation System (LCRNS) — Section 3.1.4.1
- Exploration Extra-Vehicular Activity (xEVA) Systems — Section 3.1.4.4.1
- Commercial Lunar Payload Services (CLPS) — Section 3.1.4.5.4
- Deep Space Logistics (DSL) — Section 3.1.4.3.1
- Deep Space Network (DSN) / Lunar Exploration Ground System (LEGS) — Section 3.1.4.1
- Payloads — Section 3.1.4.6.1
- Unallocated — Section 3.1.5

Table 3-2. Mapping of Use Cases to Element for Human Lunar Return Segment

Note: "X" indicates mapping of use cases to elements; It does not indicate that an element fully satisfies the use case, function, or blueprint objective, or that completion is achieved. Element descriptions can be referenced in section 3.1.4.	SLS	EGS	Orion	Gateway	HLS	LCRNS	xEVA System	CLPS	DSL	DSN / LEGS	Payloads
Transport crew and systems from Earth to cislunar space	X	X	X								
Staging of crewed lunar surface missions from cislunar space			X	X	X				X		
Aggregation and physical assembly of spacecraft elements in cislunar space	X		X	X							
Crew transport between cislunar space and lunar surface					X						
Return crew and systems from cislunar space to Earth			X								
Crew operations on lunar surface					X	X	X			X	
Frequent crew EVA on surface					X		X				
Crew conduct utilization activities on surface					X		X				
Crew conduct utilization activities in cislunar space			X	X	X				X		X
Crew emplacement and set-up of science/utilization packages on lunar surface					X		X				
Crew emergency health care and monitoring while in transit			X		X						
Return of collected samples to Earth in sealed sample containers			X		X						
Crewed missions to distributed landing sites around South Pole					X						
Crew EVA exploration and identification of samples							X				
Crew collection of samples from lighted areas							X				
Crew emplacement and set-up of science packages on lunar surface w/ long-term remote operation					X		X				
Crew emplacement and set-up of Heliophysics packages at cislunar elements w/ long-term remote operation				X					X		
Autonomous deployment and long-term operation of free-flying packages in various lunar orbits	X								X		
Crew emplacement and set-up of Heliophysics packages on lunar surface w/ long-term remote operation					X		X				
Crew collection of regolith samples from a variety of sites					X		X				

	Note: "X" indicates mapping of use cases to elements; It does not indicate that an element fully satisfies the use case, function, or blueprint objective, or that completion is achieved. Element descriptions can be referenced in section 3.1.4.	SLS	EGS	Orion	Gateway	HLS	LCRNS	xEVA System	CLPS	DSL	DSN / LEGS
1	Crew emplacement and set-up of fundamental physics experiments at cislunar elements w/ long-term remote operation				X					X	
5	Provide advanced geology training as well as detailed objective-specific training to astronauts for science activities prior to each Artemis mission										
6	Train astronauts for science tasks during an Artemis mission utilizing in situ training capabilities										
7	Allow ground personnel and science team to directly engage with astronauts on the surface and in lunar orbit, augmenting the crew's effectiveness at conducting science activities.				X	X	X	X			X
3	Crew emplacement and set-up of physics packages at cislunar elements w/ long-term remote operation				X					X	
7	Crew and/or robotic emplacement and set-up of science instrumentation in lunar orbit w/ long-term remote operation				X					X	
8	Autonomous deployment and long-term operation of free-flying packages in various lunar and heliocentric orbits	X								X	
2	Crew emplacement and set-up of physics packages at cislunar elements w/ long-term remote operation				X					X	
3	Crew emplacement and set-up of physics packages on lunar surface w/ long-term remote operation					X		X			
6	Deployment of assets in lunar orbit to provide high availability, high bandwidth communication from a variety of exploration locations on the lunar surface to Earth					X	X				X
7	Communications between assets on the lunar surface					X		X			
8	Deployment of assets in lunar orbit to provide high availability position, navigation, and timing for astronauts and robotic elements at exploration locations on the lunar surface						X				
1	Landing of crew vehicle at specific pre-defined location within exploration area				X	X					
3	Crew exploration around landing site or around habitation elements in EVA suits							X			
6	Deploy and demonstrate operation of capability to recover oxygen from lunar regolith								X		
7	Deploy and demonstrate operation of capability to recover polar water/volatiles								X		
2	Conduct autonomous/semi-autonomous mission operations in cislunar space				X						

Note: "X" indicates mapping of use cases to elements; It does not indicate that an element fully satisfies the use case, function, or blueprint objective, or that completion is achieved. Element descriptions can be referenced in section 3.1.4.	SLS	EGS	Orion	Gateway	HLS	LCRNS	xEVA System	CLPS	DSL	DSN / LEGS	Payloads
Conduct crew EVA exploration on the lunar surface					X		X				
Crew use of EVA tools to collect samples, clean suits, etc.							X				

Table 3-3. Mapping of Functions to Elements for Human Lunar Return Segment

Note: "X" indicates mapping of functions to elements; It does not indicate that an element fully satisfies the use case, function, or blueprint objective, or that completion is achieved. Element descriptions can be referenced in section 3.1.4.	SLS	EGS	Orion	Gateway	HLS	LCRNS	xEVA System	CLPS	DSL	DSN / LEGS	Payloads
Provide ground services		X									
Stack and integrate	X	X									
Manage consumables and propellant		X									
Enable vehicle launch	X	X									
Provide multiple launch attempts	X	X									
Provide aborts	X		X		X						
Transport crew and systems from Earth to cislunar space	X	X	X								
Vehicle rendezvous, proximity ops, docking, and undocking in cislunar space			X	X	X				X		
Provide PNT capability in cislunar space						X					
Provide crew habitation in cislunar space				X	X						
Transport elements from Earth to cislunar space	X										
Docking/berthing of spacecraft elements			X	X	X				X		
Transport crew and systems from cislunar space to lunar surface South Pole sites					X						
Transport crew and systems from lunar surface to cislunar space					X						
Transport crew and systems from cislunar space to Earth			X								

Note: "X" indicates mapping of functions to elements; It does not indicate that an element fully satisfies the use case, function, or blueprint objective, or that completion is achieved. Element descriptions can be referenced in section 3.1.4.	SLS	EGS	Orion	Gateway	HLS	LCRNS	xEVA System	CLPS	DSL	DSN / LEGS
Recover crew, systems, and cargo after splashdown		X								
Unload cargo on lunar surface					X					
Generate power on lunar surface					X					
Store power on lunar surface					X					
Provide high bandwidth, high availability comms between lunar surface and Earth				X	X	X				X
Provide PNT capability on the lunar surface						X				
Provide pressurized, habitable environment on lunar surface					X					
Conduct crew surface EVA activities							X			
Allow crew ingress/egress from habitable elements to vacuum					X					
Recover and package surface samples							X			
Transport cargo from Earth to elements in deep space									X	
Provide pressurized, habitable environment in deep space			X	X	X					
Move cargo into habitable elements in deep space			X	X	X				X	
Remove trash from habitable elements in deep space				X					X	
Provide crew health maintenance			X	X	X					
Transport cargo from lunar surface to Earth			X		X					
Recover samples after splashdown		X								
Crew survey of areas of interest and identification of samples							X			
Provide tools and containers to recover and package surface samples							X			
Store collected samples on lunar surface							X			
Orbital observation and sensing of lunar surface										
Deliver utilization cargo to cislunar elements				X					X	
External mounting points on cislunar elements				X						
Deliver free-flyers to cislunar space	X								X	
Aggregate and extended storage collected samples in cislunar space				X						

Note: "X" indicates mapping of functions to elements; It does not indicate that an element fully satisfies the use case, function, or blueprint objective, or that completion is achieved. Element descriptions can be referenced in section 3.1.4.	SLS	EGS	Orion	Gateway	HLS	LCRNS	xEVA System	CLPS	DSL	DSN / LEGS	Payloads
Provide training of crew prior to mission											
Provide high bandwidth, high availability comms between cislunar space and Earth				X		X				X	
Provide Earth ground stations for exploration communications										X	
Provide high bandwidth, high availability comms between lunar surface assets						X				X	
Provide precision landing system for crew transport to lunar surface					X						
Demonstrate operation of oxygen production ISRU											X
Transport autonomous payloads from Earth to lunar surface								X			
Unload autonomous payloads on lunar surface								X			
Demonstrate operation of water production ISRU											X
Provide solar observation and monitoring capability											X
Transport crew and systems from cislunar space to lunar surface					X						
Autonomous element command control in cislunar space				X					X		
Provide EVA tools to collect samples							X				
Provide EVA tools to clean EVA suits and equipment							X				

3.1.3 Reference Missions and Concepts of Operations

As described in the Objective decomposition methodology, use cases may be grouped into reference missions in order to show examples of how several use cases may be accomplished with a particular concept of operations. Table 3-2, presented previously, shows the full set of use cases in HLR, so only a representative subset is discussed below in two reference missions.

3.1.3.1 Crewed Initial Lunar Surface Reference Mission

As the first crewed mission returning to the lunar surface, this RM encompasses many use cases that will be repeated throughout the Moon-to-Mars campaign. Starting with transportation, use cases include transporting crew and systems from Earth to cislunar space, staging crewed lunar surface missions from cislunar space, assembling integrated assets in cislunar space, transporting crew and systems between cislunar space and the lunar surface, and returning crew and systems from cislunar space to Earth. The surface portion includes use cases such as crew operations on the lunar surface, frequent crew EVAs on the surface, and crew conducted utilization activities (including science, crew health and performance, and other operations) on the surface and in space.

3.1.3.2 Crewed Gateway and Lunar Surface Reference Mission

Building up from the initial return mission to the lunar surface, more capabilities in cislunar space address additional use cases, particularly for lunar orbital operations. As a habitable outpost located in NRHO, Gateway enables additional use cases in HLR beyond those in the initial crewed mission to the lunar surface. In particular, Gateway allows for crew to conduct utilization activities in cislunar space; allows for ground personnel and science teams to directly engage with astronauts on the surface and in lunar orbit, augmenting the crew's effectiveness at conducting science activities; enables crew and/or robotic emplacement and set-up of science instrumentation in lunar orbit with long-term remote operation; and includes autonomous/semi-autonomous mission operations in cislunar space.

3.1.4 Element Descriptions

Elements represent capabilities that are available in the HLR campaign segment that meet the designated Agency objectives and derived functions needed to support those objectives. The elements are described in the sub-architectures they support and are not in chronological order.

3.1.4.1 Communication, Positioning, Navigation, and Timing

During the HLR, CPNT services will be provided through a combination of assets on Earth, in lunar orbit, and on the lunar surface. Direct-to-Earth (DTE) service needs will be met through a combination of an upgraded Deep Space Network (DSN) as well as a dedicated new Lunar Exploration Ground System (LEGS) subnet. Three initial LEGS sites around Earth will provide continuous coverage of the Moon with the option for commercial services and international partner support to augment capacity and redundancy. Orbiting assets such as Gateway, the Lunar Communications Relay and Navigation System (LCRNS), and possible partner assets will provide service to users without line-of-sight to Earth as well as reducing the required size, weight, and power for a user's communications systems. The LCRNS will initially, in this segment, cover a service volume from -80° S to the South Pole of the Moon and up to 125 km altitude. Communication will support one S-band bidirectional link and one simultaneous Ka-band return link as well as CPNT service through an Augmented Forward Signal (AFS). In the later part of the HLR segment, LCRNS service will expand to bidirectional Ka-band and multiple AFS links.

Surface-to-surface communications may initially rely on legacy systems such as Ultra High Frequency (UHF) and WiFi but will seek to leverage terrestrial standards such as 3GPP/5G within this segment of the architecture to increase mobility and capacity. Throughout this initial phase of the architecture, interoperability will be emphasized through the LunaNet Interoperability Specification (LNIS), the International Communication System Interoperability Standards (ICSIS)[7], and similar specifications. The growth of CPNT services throughout the HLR segment will enable the near-term exploration objectives of the HLR segment while providing a robust foundation upon which a scalable infrastructure can grow to support the needs of a sustained lunar presence, including precursor missions that will inform and validate a Martian architecture.

The functions the LCRNS fulfills in the HLR campaign segment are shown in Table 3-4.

Table 3-4. Functions Fulfilled by LCRNS

F-09	Provide PNT capability in cislunar space
F-23	Provide high bandwidth, high availability comms between lunar surface and Earth
F-24	Provide PNT capability on the lunar surface
F-71	Provide high bandwidth, high availability comms between cislunar space and Earth
F-82	Provide high bandwidth, high availability comms between lunar surface assets

The functions the DSN and LEGS fulfill in the HLR campaign segment are shown in Table 3-5.

Table 3-5. Functions Fulfilled by DSN/LEGS

F-23	Provide high bandwidth, high availability comms between lunar surface and Earth
F-78	Provide Earth ground stations for exploration communications
F-81	Provide Earth ground stations for exploration communications
F-82	Provide high bandwidth, high availability comms between lunar surface assets

3.1.4.2 Habitation

3.1.4.2.1 Gateway[8] Crew Capable Configuration Overview

The Gateway architecture is composed of several modules incrementally launched and assembled in NRHO around the Moon in a system that provides for continuous architectural evolution. Individual Gateway modules are launched either as co-manifested payloads (CPL) on the Space Launch System (SLS) along with the Orion crew vehicle or on commercial launch vehicles. The modules combined in the Gateway architecture represent a meaningful series of demonstration steps in the direction of enabling the more extensive exploration effort in the future.

The HLR campaign segment comprises the Gateway Crew Capable Configuration: Power and Propulsion Element (PPE), Habitation and Logistics Outpost (HALO), International Habitation Module (I-HAB) and Gateway Logistics Module. For this segment, Gateway capability represents a minimum functional core to support the initial human landing missions to the lunar surface. The I-HAB is being provided by ESA with contributions from the Japan Aerospace Exploration Agency (JAXA). These modules provide pressurized volume for the crew to move between the docked vehicles, crew habitation activities (food and water consumption, sleep, hygiene), and internal and external utilization capabilities. They also provide initial life support services and docking ports for additional modules and visiting vehicles. The PPE is a commercially based spacecraft that provides electrical power, attitude and translational control, and communication for the Gateway. The PPE maintains attitude through the use of reaction wheels and a chemical

---

[7] International Deep Space Interoperability Standard. www.internationaldeepspacestandards.com
[8] For more information, please visit: www.nasa.gov/gateway

propulsion system. Translation maneuvers and orbital maintenance are primarily performed using a Solar Electric Propulsion (SEP) system. The PPE has power storage and the systems necessary to convert and distribute power to the rest of the Gateway. It provides internal avionics systems and is one part of an integrated command and control architecture for Gateway.

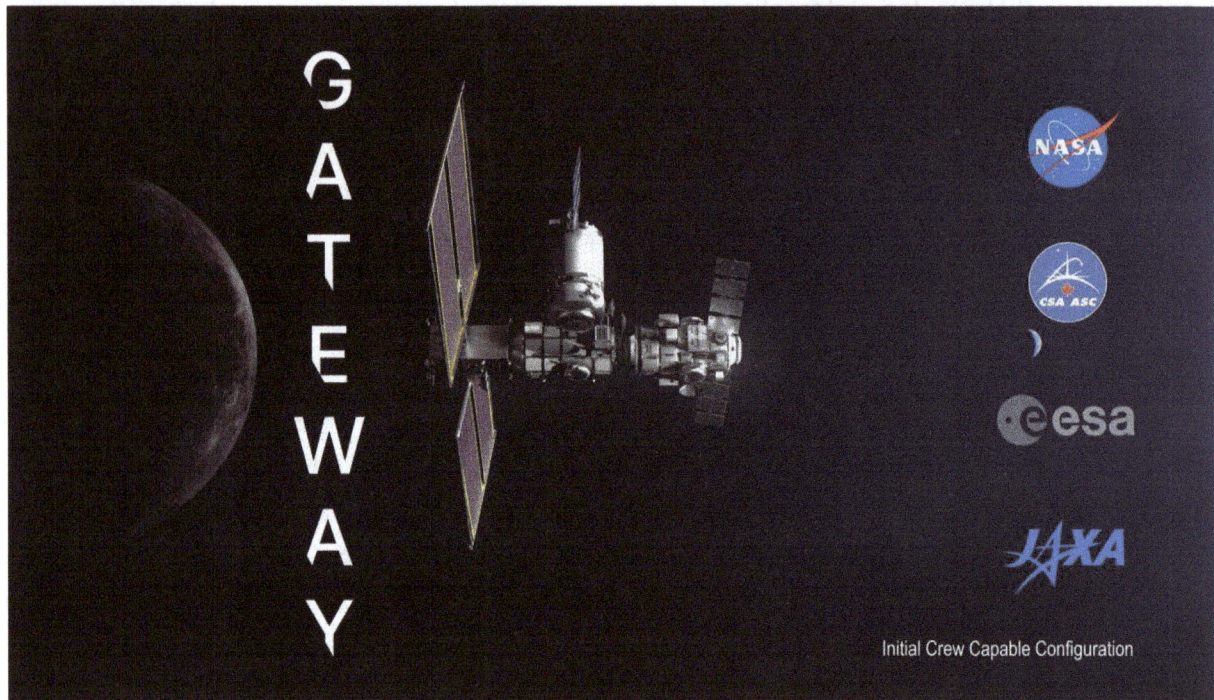

Figure 3-1. Gateway Crew Capable Configuration

The integrated PPE/HALO configuration also provides communication via PPE and the ESA HALO Lunar Communications Systems (HLCS) for space-to-Earth and space element to space element; with visiting vehicles during rendezvous, proximity operations, and docking/undocking; and between lunar surface systems and Earth. NASA utilizes Deep Space Logistics (see Section 3.1.4.3) to deliver cargo and other supplies to Gateway, including critical spares and outfitting for HALO and I-HAB, cargo stowage, and trash disposal. Gateway will launch with an initial suite of internal and external science utilization payloads, provided by NASA, ESA, and JAXA, that will operate and collect data in transit and in NRHO during crewed and uncrewed operations. External payload sites and future robotic attach points will be provided by CSA on PPE, HALO, and I-HAB. The Gateway Crew Capable Configuration is shown in Figure 3-1. Expansion of the Gateway is planned to include additional capabilities and systems as part of the Foundational Exploration segment.

The functions Gateway Crew Capable Configuration fulfills in the HLR campaign segment are shown in Table 3-6.

Table 3-6. Functions Fulfilled by Gateway Crew Capable Configuration

F-08	Vehicle rendezvous, proximity operations, docking, and undocking in cislunar space
F-10	Provide crew habitation in cislunar space
F-12	Docking/berthing of spacecraft elements
F-23	Provide high bandwidth, high availability comms between lunar surface and Earth
F-34	Provide pressurized, habitable environment in deep space
F-35	Move cargo into habitable elements in deep space

F-36	Remove trash from habitable elements in deep space
F-38	Provide crew health maintenance
F-56	Deliver utilization cargo to cislunar elements
F-57	External mounting points on cislunar elements
F-63	Aggregate and extended storage collected samples in cislunar space
F-71	Provide high bandwidth, high availability comms between cislunar space and Earth
F-100	Autonomous element command control in cislunar space

3.1.4.3　Logistics Systems

3.1.4.3.1　Cislunar Logistics Overview

Exploration activities will need logistics deliveries to support the achievement of objectives. Logistics represents all equipment and supplies that are needed to support mission activities that are not installed as part of the vehicle. Logistics typically includes consumables (e.g., food, water, oxygen), maintenance items (planned replacement items), spares (for unexpected/unplanned failures), utilization (e.g., science and technology demonstrations), and outfitting (additional systems/sub-systems for the elements), as well as the associated packaging. Logistics deliveries of critical pressurized and unpressurized cargo and payloads will be needed to support activities with and without crew. In the HLR segment of the exploration campaign, logistics delivery will be provided by NASA's Deep Space Logistics (DSL)[9].

Figure 3-2. Dragon XL Cislunar Logistic Module

During HLR, DSL is responsible for leading the commercial supply chain in deep space by procuring services for transporting cargo, payloads, equipment, and consumables to enable exploration of the Moon and Mars. Logistics flights are necessary to supply Gateway with critical cargo deliveries and maximize the length of crew stays on Gateway. The Gateway Logistics Services contract and technical capability are extensible to deliver unique payload configurations and supply cargo deliveries to other destinations. Additional capabilities may be added in future segments. At least one logistics services delivery is anticipated for each Artemis mission to

---

[9] For more information, please visit: www.nasa.gov/content/about-gateway-deep-space-logistics

Gateway of 30 days. Dragon XL is shown in Figure 3-2 as one of the providers of Gateway logistics.

The functions the DLS fulfills in the HLR campaign segment are shown in Table 3-7.

Table 3-7. Functions Fulfilled by Deep Space Logistics

F-08	Vehicle rendezvous, proximity ops, docking, and undocking in cislunar space
F-12	Docking/berthing of spacecraft elements
F-33	Transport cargo from Earth to elements in deep space
F-35	Move cargo into habitable elements in deep space
F-36	Remove trash from habitable elements in deep space
F-56	Deliver utilization cargo to cislunar elements
F-58	Deliver free-flyers to cislunar space

3.1.4.4 Mobility Systems

3.1.4.4.1 Exploration Extra-Vehicular Activity (xEVA) System Overview

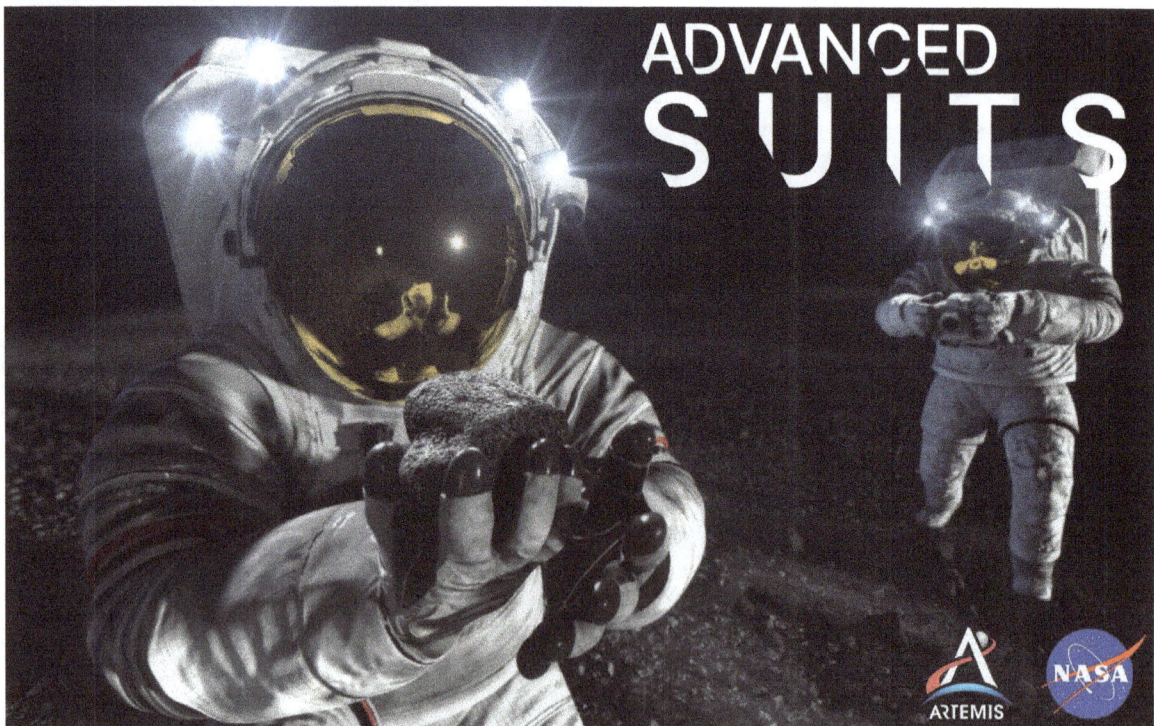

Figure 3-3. Exploration Extravehicular Activity System

The xEVA System allows crew members to perform extravehicular exploration, research, construction, servicing, repair operations, and utilization and science in cislunar orbit and on the lunar surface. EVA transverse and tasks may be augmented by robotics and rovers. The xEVA System includes the EVA suit, EVA tools, and vehicle interface equipment. Figure 3-3 provides an overview of the xEVA System. Through Exploration Extravehicular Activity Services, Axiom Space and Collins Aerospace have been selected to build the next generation of spacesuit and spacewalk systems.

The functions the xEVA fulfills in the HLR campaign segment are shown in Table 3-8.

Table 3-8. Functions Fulfilled by xEVA System

F-28	Conduct crew surface EVA activities
F-32	Recover and package surface samples
F-46	Crew survey of areas of interest and identification of samples
F-47	Provide tools and containers to recover and package surface samples
F-48	Store collected samples on lunar surface
F-105	Provide EVA tools to collect samples
F-106	Provide EVA tools to clean EVA suits and equipment

3.1.4.5 Transportation

3.1.4.5.1 Space Launch System (SLS)[10] Overview

Figure 3-4. Space Launch System

The SLS is a super-heavy-lift launch vehicle that provides the foundation for human exploration beyond Earth orbit (BEO). With its unprecedented power and capabilities, SLS is the only launch vehicle that can send Orion, astronauts, and payloads directly to the Moon on a single launch. The SLS is designed to be evolvable, which makes it possible to conduct more types of missions, including human missions to Mars; assembly of large structures; and robotic, scientific, and exploration missions to destinations such as the Moon, Mars, Saturn, and Jupiter. Humans will be transported safely, and different payloads will be delivered efficiently and effectively, to enable a variety of complex missions in cislunar and deep space. The first SLS crew transportation system, called Block 1, uses an Interim Cryogenic Propulsion Stage (iCPS) to send the Orion spacecraft on towards the Moon. Block 1 was used for Artemis I and is planned for use for Artemis

---

[10] For more information, please visit: www.nasa.gov/exploration/systems/sls/index.html

II and III. The Block 1B variant will use an Exploration Upper Stage (EUS) to enable more ambitious missions, such as carrying the Orion crew vehicle along with large cargo (co-manifested payload) in a single launch. SLS also enables free flyer science payloads in cislunar and beyond as secondary payloads. Although Block 1 and Block 1B Crew are the only two variants in HLR, Block 1B Cargo and Block 2 Crew and Cargo variants are key capabilities for future campaign segments. Figure 3-4 exhibits the SLS in the Block 1 configuration as flown for Artemis I.

The functions the SLS fulfills in the HLR campaign segment are shown in Table 3-9.

Table 3-9. Functions Fulfilled by SLS

F-02	Stack and integrate
F-04	Enable vehicle launch
F-05	Provide multiple launch attempts
F-06	Provide aborts
F-07	Transport crew and systems from Earth to cislunar space
F-11	Transport elements from Earth to cislunar space
F-58	Deliver free-flyers to cislunar space

3.1.4.5.2 Orion[11] Overview

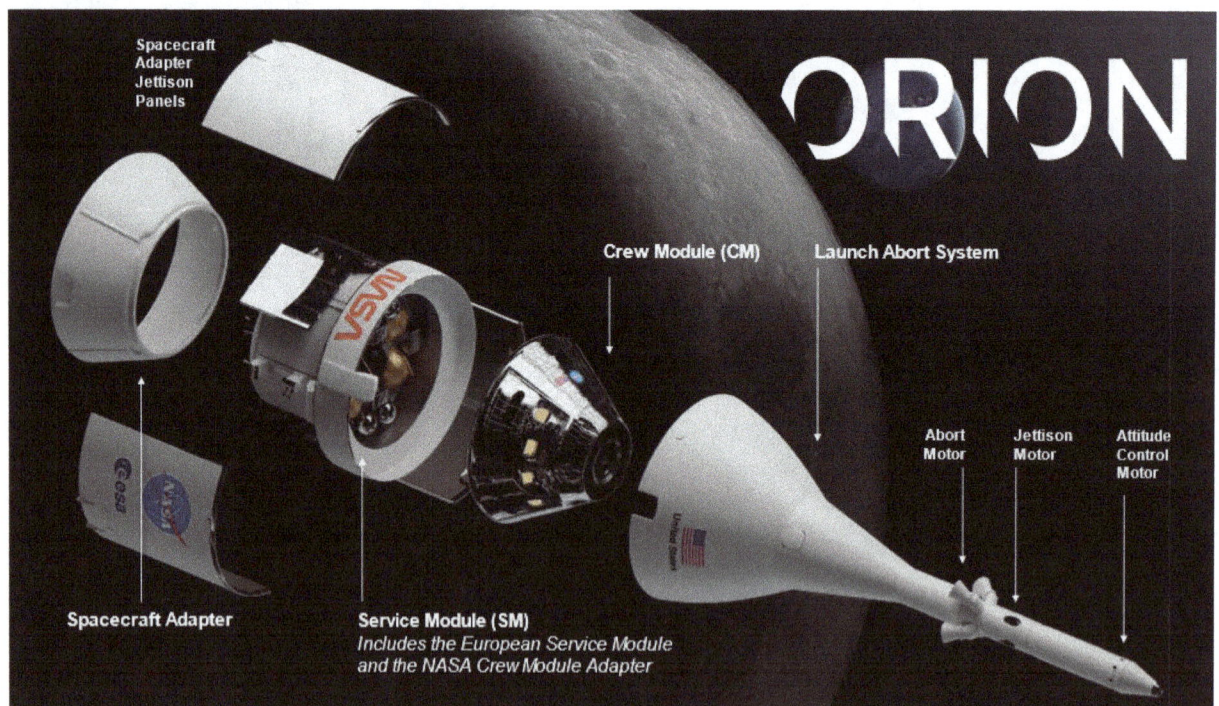

Figure 3-5. Orion Spacecraft

The Orion spacecraft, NASA's next-generation spaceship to take astronauts on a journey of exploration to the Moon and on to Mars, is shown in Figure 3-5. The Orion spacecraft serves as the primary crew vehicle for Artemis missions for transporting crew between Earth and lunar orbit. The vehicle can conduct regular in-space operations in conjunction with payloads delivered by the SLS. The Orion spacecraft includes the Crew Module (CM), Service Module (SM), and Launch Abort System (LAS). The CM is capable of transporting four crew members beyond the Moon,

---

[11] For more information, please visit: www.nasa.gov/exploration/systems/orion/index.html

providing a safe habitat from launch through landing and recovery. The SM, made up of the NASA-provided Crew Module Adapter (CMA) and the ESA-provided European Service Module (ESM), provides support to the crew module from launch through separation prior to entry. The SM provides in-space propulsion for orbital transfer, power and thermal control, attitude control and high-altitude ascent aborts. While mated with the crew module, the SM also provides water and air to support the crew. The LAS, positioned on a tower atop the CM, can activate within milliseconds to propel the vehicle to safety and position the CM for a safe landing.

The functions the Orion spacecraft fulfills in the HLR campaign segment are shown in Table 3-10.

Table 3-10. Functions Fulfilled by Orion

F-06	Provide aborts
F-07	Transport crew and systems from Earth to cislunar space
F-08	Vehicle rendezvous, proximity operations, docking, and undocking in cislunar space
F-10	Provide crew habitation in cislunar space
F-12	Docking/berthing of spacecraft elements
F-16	Transport crew and systems from cislunar space to Earth
F-34	Provide pressurized, habitable environment in deep space
F-35	Move cargo into habitable elements in deep space
F-38	Provide crew health maintenance
F-42	Transport cargo from lunar surface to Earth

3.1.4.5.3 Human Landing System (HLS) —Initial and Integrated Lander Configurations Overview

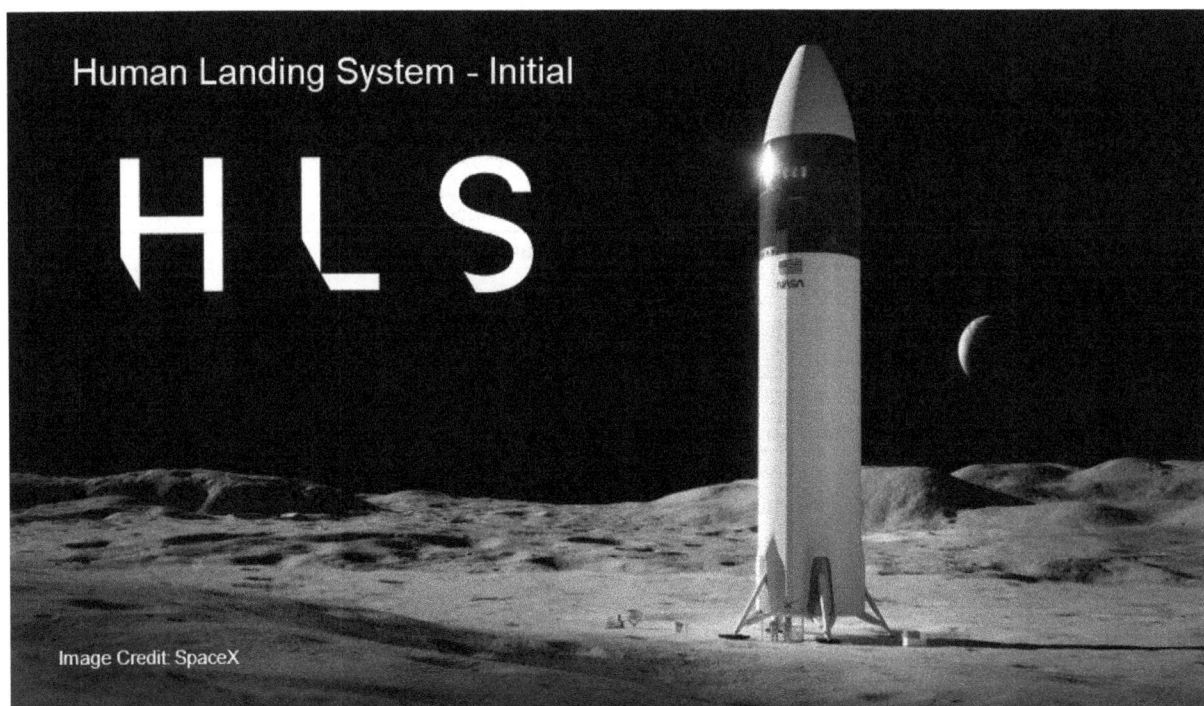

Figure 3-6. Human Landing System—Initial as Awarded

The Human Landing System (HLS) will transport crew members, support payloads, cargo, and logistics between a crew staging vehicle (either Orion or Gateway) orbiting the Moon in NRHO and the lunar surface. On the lunar surface, HLS provides the habitable volume, consumables, and design features enabling crew surface stay and execution of lunar surface EVAs, along with utilization accommodations inside the cabin as well as external attached payloads. The specific HLS architecture is subject to commercial provider design implementation approach.

The initial HLS configuration supports a crew of two and will operate between Orion in NRHO and a landing site in the vicinity of the lunar South Pole. Additionally, in this configuration HLS will deliver the cargo and support logistics to NRHO from Earth prior to the start of the crewed phase of the mission. The initial human landing mission will be a demonstration of this initial HLS configuration and of the minimum basic technologies and innovation required to safely transport crew and utilization cargo to and from the lunar surface. The initial HLS is shown in Figure 3-6.

The HLS integrated lander will build on the initial configuration's base capabilities to enable the full range of crewed lunar mission objectives, including accommodating additional internal and external payloads. More ambitious missions will also be pursued as lunar surface exploration evolves towards the Foundational Exploration segment. Missions with the HLS integrated lander will require HLS to support landing a crew of up to 4, leveraging additional habitable surface assets to support the larger crew for the duration of the lunar stay. These missions may include the capability to land and operate at non-polar landing sites or for extended durations at the lunar south pole. This HLS configuration has increased performance capabilities allowing for enhanced up and down mass and increased darkness survivability. These missions will also seek sustainable HLS designs that may include reusable elements or interactions with other systems in the lunar vicinity. All missions with the HLS integrated lander will begin and end at the Gateway—enabling extended missions on the lunar surface as Orion will be able to remain in lunar orbit longer docked to the Gateway.

The functions the HLS fulfills in the HLR campaign segment are shown in Table 3-11.

Table 3-11. Functions Fulfilled by HLS

F-06	Provide aborts
F-08	Vehicle rendezvous, proximity operations, docking, and undocking in cislunar space
F-12	Docking/berthing of spacecraft elements
F-13	Transport crew and systems from cislunar space to lunar surface South Pole sites
F-14	Transport crew and systems from lunar surface to cislunar space
F-19	Unload cargo on lunar surface
F-21	Generate power on lunar surface
F-22	Store power on lunar surface
F-23	Provide high bandwidth, high availability comms between lunar surface and Earth
F-25	Provide pressurized, habitable environment on lunar surface
F-29	Allow crew ingress/egress from habitable elements to vacuum
F-34	Provide pressurized, habitable environment in deep space
F-35	Move cargo into habitable elements in deep space
F-38	Provide crew health maintenance
F-42	Transport cargo from lunar surface to Earth
F-87	Provide precision landing system for crew transport to lunar surface
F-99	Transport crew and systems from cislunar space to lunar surface

3.1.4.5.4 Cargo Transportation—Landers—Commercial Lunar Payload Services (CLPS)[12]

Lunar surface exploration will require the delivery of assets, equipment, and supplies to the lunar surface. While some supplies and equipment may be delivered with crew on HLS, cargo landers

---

[12] For more information, please visit www.nasa.gov/clps.

provide additional flexibility and capability for robust exploration. In the HLR segment of the exploration campaign, additional cargo delivery can be provided through NASA's CLPS.

NASA's CLPS initiative allows rapid acquisition of lunar delivery services from American companies for payloads that advance capabilities for science, technology, exploration, or commercial development of the Moon. Investigations and demonstrations launched on commercial Moon flights will help the Agency study Earth's nearest neighbor under the Artemis approach. Companies are encouraged to fly commercial and other partner payloads in addition to the NASA payloads. NASA has awarded 9 task orders to CLPS lander providers for delivery of more than 40 payloads to the lunar surface during the HLR exploration segment. Additional task orders will be awarded as mission and payload definition continues. Current CLPS deliveries are sending science and technology payloads. NASA's Science Mission Directorate (SMD) is planning annual calls for new payload suites called the Payload and Research Investigations from the Surface of the Moon (PRISM). PRISM will enable high priority science and will be complemented by other NASA-sponsored payloads.

The functions CLPS fulfills in the HLR campaign segment are shown in Table 3-12.

Table 3-12. Functions Fulfilled by CLPS

F-91	Transport autonomous payloads from Earth to lunar surface
F-92	Unload autonomous payloads on lunar surface

3.1.4.6 Utilization Systems

3.1.4.6.1 Payloads Overview

The transportation, delivery, deployment, and operation of utilization payloads to cislunar space and the lunar surface, as well as the return to Earth of samples and other cargo, is a key service provided by the Moon-to-Mars Architecture and a critical enabler of every NASA utilization objective. Utilization payload is broadly defined here to encompass any item transported and supported by the M2M Architecture that is primarily in support of and attributed to utilization objectives, as distinct from other components in the baseline platform of services provided by the Architecture. Utilization payload includes internal and external equipment, scientific experiments, technology demonstrators, instruments, tools, supplies, containers, and samples transported by any crewed or robotic element in the Architecture. Examples include:

- Secondary SLS payloads, including CubeSats
- Externally mounted scientific sensors on Gateway, HLS, logistics modules, and other surface elements
- Science experiments and technology demonstrators deployed to the lunar surface by the crew or by robotic landers
- Internally operated experiments and other equipment in every crew volume, including Orion, Gateway, and HLS
- Tools and containers used to collect geological samples from the lunar surface, as well as samples collected from other science experiments and human research activities
- Portable equipment used to make scientific observations of the lunar surface, including cameras and other instruments

Note that some equipment, including some multi-purpose cameras and medical equipment, is dual-use in supporting both utilization and operations, and may be considered a utilization payload or a part of the platform depending on the context.

The functions the payloads fulfill in the HLR campaign segment are shown in Table 3-13.

Table 3-13. Function Fulfilled by Payloads

F-50	Orbital observation and sensing of lunar surface
F-90	Demonstrate operation of oxygen production ISRU
F-93	Demonstrate operation of water production ISRU
F-97	Provide solar observation and monitoring capability

3.1.4.7 Other Elements

3.1.4.7.1 Exploration Ground Systems (EGS)[13] Overview

Figure 3-7. Exploration Ground Systems

EGS was established to develop and operate systems and facilities necessary to process, launch, and recover vehicles. The EGS Program provides the ground infrastructure for launch and landing in support of processing and launch of the SLS and Orion. EGS also provides recovery capabilities of the Orion spacecraft. EGS utilizes the Vehicle Assembly Building (VAB) for integration and testing, as well as vertical stacking on the Mobile Launcher (ML). The ML with the SLS and Orion secured is moved to Launch Pad 39B by the crawler-transporter. Vehicle testing, vehicle final propellant servicing, launch countdown, and launch take place at launch pad 39B. Additional capabilities, such as the Mobile Launcher 2 (ML2) to support the SLS Block 1B, will be included in the infrastructure of EGS to support future missions. The VAB is shown in Figure 3-7.

The functions EGS fulfills in the HLR campaign segment are shown in Table 3-14.

.

---

[13] For more information, please visit: www.nasa.gov/exploration/systems/ground/index.html

Table 3-14. Functions Fulfilled by EGS

F-01	Provide ground services
F-02	Stack and integrate
F-03	Manage consumables and propellant
F-04	Enable vehicle launch
F-05	Provide multiple launch attempts
F-07	Transport crew and systems from Earth to cislunar space
F-17	Recover crew, systems, and cargo after splashdown
F-43	Recover samples after splashdown

3.1.5 Unallocated Use Cases and Functions

Use case and functional decomposition has been completed and focused on near-term achievability of the lunar objectives. An assessment needs to be completed of the Foundational Exploration objectives to determine if any of those use case and functions should be brought into the HLR segment. In addition, once the Mars objectives decomposition is complete, there may be additional lunar use cases and functions identified to be included in the HLR segment.

There are two use cases (UC-45, UC-46) and one function (F-68) that are unallocated. This likely will be allocated to the "human" element as that sub-architecture is further refined.

- UC-45: Provide advanced geology training as well as detailed objective-specific training to astronauts for science activities prior to each Artemis mission
- UC-46: Train astronauts for science tasks during an Artemis mission utilizing in situ training capabilities
- F-68: Provide training of crew prior to mission

3.1.6 Open Questions, Ongoing Assessments, and Future Work

Open questions, on-going assessments, and future work for HLR segment include:

- What options are available to increase sample return and conditioned cargo from the lunar surface to Earth?
- What options are available to increase down-mass to the lunar surface to support utilization?
- What implications to the current systems are required to support non-polar sorties in the HLR segment?
- What elements need to provide in situ training of crew in cislunar space?

3.2 FOUNDATIONAL EXPLORATION SEGMENT

The Foundational Exploration (FE) segment builds on the initial capabilities of Human Lunar Return and prepares for future segments through the lunar expansion of operations, capabilities, and systems supporting complex orbital and surface missions to conduct utilization (science, etc.) and Mars-forward precursor missions. With the continued usage of the elements in HLR and the deployment of new capabilities, surface missions will feature increased duration, expanded mobility, and regional exploration of the lunar South Pole. Orbital operations will also increase in duration; these, when coupled with the surface mission phases, will serve as Mars mission analogs, validating both the systems and the exploration concept of operations for future Mars mission profiles. FE will by necessity have to initiate activities and capabilities that will be influenced by the future needs in the SLE and Humans to Mars segment. Such activities include reconnaissance, Mars risk reduction, and initial infrastructure supporting the long-term SLE evolution.

3.2.1 Summary of Objectives

Increased mission durations, expanded capabilities, and the ability to access additional regions of the lunar surface enable a growth in utilization, during both crewed and uncrewed mission phases. A variety of science objectives may be addressed during the FE segment, ranging from lunar and planetary science to human and biological science and including science-enabling and applied science goals. During the FE campaign segment, enhanced architecture capabilities would further enhance ability to address and achieve science objectives, including 1) expanding accessible regions of exploration from the south polar region to key locations across the Moon to further advance understanding of the chronology, composition, and internal structure of the Moon (LPS-1 and LPS-2), 2) characterizing the distribution, source, and composition of volatile-bearing materials across the lunar south polar region, including within larger permanently shadowed regions (LPS-3) and determine their viability for ISRU, 3) generating forecasting capabilities for space weather monitoring off the Earth-Sun line (HS-1), 4) characterizing plant, model organisms/systems, and human physiological responses exposed long-term to extreme environments with micro- and 1/6-gravitational strengths (HBS-1, HBS-3), 5) characterizing physical systems in 1/6-gravity environments and associated models (HBS-2), and 6) testing the theory of relativity and quantum physics in the lunar environment (PPS-1, PPS2). These FE science priorities were identified by the Science Mission Directorate.

The majority of the Lunar Infrastructure (LI) objectives help define FE (LI-4 and LI-7 will apply to future segments). Expansion of the power (LI-1), communications/position/navigation/timing (LI-2, LI-3), transportation (LI-5, LI-6), mobility (LI-6), infrastructure (LI-8), and utilization (LI-9) sub-architectures all build toward the LI goal of "[creating]   an interoperable global lunar utilization infrastructure where U.S. industry and international partners can maintain continuous robotic and human presence on the lunar surface for a robust lunar economy without NASA as the sole user, while accomplishing science objectives and testing for Mars."

The Transportation and Habitation (TH) objectives drive the additional capabilities in mobile, habitation, and transportation systems during FE. For example, TH-1, TH-2, and TH-11 all address a need for transportation systems to transfer crew and cargo to-and-from Earth, through cislunar space, and to and from lunar orbit to the surface, enabling scientific and utilization objectives. TH-3 (develop system(s) to allow crew to explore, operate, and live on the lunar surface and in lunar orbit with scalability to continuous presence; conducting scientific and industrial utilization as well as Mars analog activities) and TH-4 (develop in-space and surface habitation system(s) for crew to live in deep space for extended duration, enabling future missions to Mars) define Foundational Exploration as a campaign segment.

A number of Operations (OP) objectives prompt the capabilities needed for FE. The overall operations goal is to "conduct human missions on the surface and around the Moon followed by missions to Mars. Using a gradual build-up approach, these missions will demonstrate technologies and operations to live and work on a planetary surface other than Earth, with a safe return to Earth at the completion of the missions." These objectives encompass the need for extended duration missions in deep space and partial-gravity environments to test systems and crew concepts of operations in preparation for the initial human Mars exploration campaign (OP-1, OP-2, OP-4, OP-5, OP-6, OP-7). Additionally, the need to develop methods to work with robotic systems (OP-9, OP-10) and characterize in-situ resources (OP-3) defines other aspects of FE.

3.2.2 Use Cases and Functions

As seen in the HLR segment, by starting with the Agency objectives and their associated characteristics and needs, particular use cases and functions necessary to accomplish the use cases may be defined. As the Foundational Exploration segment matures, so will the functional breakdown from the objectives.

As a representative example, objective TH-3 (develop system(s) to allow crew to explore, operate, and live on the lunar surface and in lunar orbit with scalability to continuous presence; conducting scientific and industrial utilization as well as Mars analog activities) drives several characteristics and needs. These include demonstration of capabilities to allow crew to live, to conduct science and utilization activities, and to exit habitable space and conduct EVA activities all in both cislunar space and on the lunar surface. Sample use cases that contribute to fulfilling those characteristics and needs include crew operations, habitation, EVA, collection of samples, and crew emplacement and set-up of science and utilization packages. Some of the functions that map to these use cases include transportation, crew health and human performance, habitation, and integrated human robotic operations.

3.2.2.1 Foundational Exploration Notional Use Cases and Functions

Although the decomposition of the use cases and functions is described in more detail in Appendix A, a representative, non-comprehensive list of use cases and key functions is provided here, specific to Foundational Exploration. These use cases and functions outline options for achieving the Agency objectives and should not be taken as a set of comprehensive plans or requirements. It is forward work to map functions to specific elements as the architecture matures and elements are defined for FE.

3.2.2.1.1.1 Foundational Exploration Notional Use Cases:

- Delivery of large elements to and unloading of elements on the lunar surface
- Robotic assistance of crew exploration, surveying sites, locating samples and resources, and retrieval of samples
- Crewed missions to landing sites around the lunar South Pole
- Crewed missions to non-polar landing sites, including the lunar far side
- Crewed/robotic collection of samples from Permanently Shadowed Regions (PSRs)
- Crew Intra-Vehicular Activities (IVA) research in dedicated science laboratory on the lunar surface
- Mars analog missions with extended durations in NRHO followed by lunar surface missions
- Deployment of power generation and storage systems at multiple locations around the lunar South Pole
- Deployment of assets in lunar orbit to provide high availability position, navigation, and timing for astronauts and robotic elements at exploration locations on the lunar surface
- Uncrewed relocation of mobility elements to landing sites around the lunar South Pole
- Crew habitation in habitable elements on surface
- Crew conduct utilization activities on Surface
- Crew excursions to locations distributed around landing site

- Crew excursions to PSRs near landing site
- Crew driving of mobility systems in EVA suits
- Crew operation of mobility systems in shirt sleeve environment
- Analog missions with extended durations in NRHO, followed by lunar surface missions
- Crew exploration around landing site or around habitation elements in EVA suits
- Crew relocation and exploration in a shirtsleeve environment

3.2.2.1.1.2 Foundational Exploration Notional Functions:

- Transport cargo from Earth to the lunar surface
- Recover and package surface samples in PSRs
- Provide pressurized, habitable environment in deep space
- Provide robotic systems on the lunar surface controlled from Earth and/or cislunar space
- Operate mobility asset in dormancy/remote mode between crewed missions
- Provide PNT capability on the lunar surface
- Transport cargo from the lunar surface to Earth
- Transport crew and systems from cislunar space to lunar surface South Pole sites
- Provide pressurized, habitable environment on lunar surface
- Move cargo into habitable elements on lunar surface
- Provide local unpressurized crew surface mobility
- Provide pressurized crew surface mobility
- Provide local unpressurized crew surface mobility into PSRs

3.2.3 Reference Missions and Concepts of Operations

As described in the Objective Decomposition methodology, use cases may be grouped into reference missions in order to show examples of how several use cases may be accomplished with a particular concept of operations. Expanding on the types of mission phases expected in HLR, several notional reference mission phases are presented to show the expansion toward achieving the objectives in Foundational Exploration.

3.2.3.1 Sortie Reference Mission with Unpressurized Mobility

The Foundational Exploration segment will build on the types of lunar surface exploration accomplished in the HLR segment, which includes crew habitation in the crew lander vehicle with capability for EVA from the lander. In FE, additional use cases may be implemented with the addition of an unpressurized mobility platform to extend EVA range and scientific exploration. This enables the use case for crew excursions to locations distributed around the landing site and has the potential to enable others such as robotic assistance of crew exploration, the locating of samples and resources, and retrieval of samples; crewed/robotic collection of samples from PSRs; and deployment of power generation, storage, and distribution systems at multiple locations around the lunar South Pole, among others.

3.2.3.2 Sortie Reference Mission with Pressurized Mobility

Working toward the objectives on expanding exploration for longer durations while conducting scientific and industrial utilization, developing surface habitation systems, and performing Mars risk reduction activities prompts the inclusion of additional functional capabilities. With initial surface crew sizes, one method to accomplish these objectives is by adding functionality for pressurized mobility systems. This function may enable use cases such as crew IVA research, additional robotic assistance of crew exploration beyond the unpressurized mobility function, expanded durations for crew operations on the lunar surface (including additional habitation functions), crew excursions to locations distributed around the landing site, EVA egress/ingress, crew/robotic collection of samples from PSRs, and crew relocation and exploration in a shirtsleeve environment.

3.2.3.3 Robotic Uncrewed Operations

Even with the opportunity to extend surface mission durations from those in HLR, the surface of the Moon will be uncrewed for the majority of each year in the Foundational Exploration segment. Functions regarding autonomous or remote operations provide the chance for additional exploration and utilization during the uncrewed portions of the year. Assuming a main function of autonomous and/or tele-operations, these robotic functions could include cargo unloading, logistics transfer, surface sample collection, and infrastructure development (e.g., landing site scouting or preparation). These functions contribute to use cases like robotic survey of potential crewed landing sites to identify locations of interest (including nearby PSRs), uncrewed relocation of mobility elements to landing sites around the lunar South Pole, and autonomous deployment of science and utilization packages.

3.2.3.4 Cislunar Operations at Gateway

A key aspect of Foundational Exploration is preparing for crewed exploration of Mars through lunar precursor missions. In addition to a growth in duration for surface mission segments from HLR, other main characteristics are to provide numerous long-duration crew increments in cislunar space prior to surface mission segments and to conduct crew transitions from microgravity to partial gravity. Extended mission segments in cislunar space at Gateway and accompanying visiting vehicles also allow for increased time for IVA science and utilization. A main use case to accomplish these characteristics and needs is to utilize precursor Mars mission profiles with extended durations in NRHO followed by lunar surface missions. Other use cases applicable to Gateway reference missions include crew delivery and transfer to crewed landing systems in cislunar space, remote diagnosis and treatment of crew health during extended increments in cislunar space, crew emplacement and setup of science and utilization packages in cislunar space (with long-term remote operation as applicable), and crew IVA research in dedicated science workspace in cislunar space.

3.2.3.5 Extended Surface Habitation Operations

The addition of dedicated surface habitation enables longer duration missions, increased crew size, and enhanced surface utilization and exploration helping to meet objectives that lead to continuous presence. With dedicated habitation capability, additional use cases to support science and utilization are achievable, enhancing crew EVA exploration, sample collection, and emplacement of science and/or utilization packages. Dedicated science workspace on the lunar surface and demonstration of bioregenerative oxygen and water recovery subsystems into the ECLS Systems in cislunar elements use cases may also be met with an additional surface habitation capability. Functions of providing longer-duration deep space, partial-gravity crew habitation, including crew medical systems and health kits; enabling surface EVA; and providing common interfaces (e.g. fluids, gases, logistics transfer, power), among many others, contribute to fulfilling the objectives.

3.2.3.6 Non-Polar Lunar Sortie Reference Mission

Although the focus for lunar surface exploration has been on the South Pole, several objectives, particularly those related to science and utilization, motivate looking at landing sites beyond the South Pole. The use case of crewed missions to non-polar landing sites would allow for exploration of alternative locations with enabling functions like crew descent, landing, and ascent at non-polar sites.

3.2.4 Notional Element and Functional Descriptions

Elements introduced in HLR will continue to be utilized, and additional capabilities will become available, flowing from the Agency objectives. These new capabilities in Foundational Exploration, although not formally assigned to defined elements at this point, can be grouped into general functional categories or sub-architectures. As the architecture matures and the Artemis program advances, new elements will be set to meet these needs. Interoperability between elements, the associated functions necessary to achieve interoperability, and the impacts of functional groupings to the overall architecture are other important aspects to consider beyond the particular element functions. Images shown are examples of concepts that may meet (or partially meet) the capabilities in these functional descriptions; they should not be taken as recommendations for design solutions or treated as the only concept(s) under consideration.

3.2.4.1 Unpressurized Mobility

Figure 3-8. Example Concepts for Unpressurized Mobility

In order to move beyond the surface mission types in HLR, new capabilities in mobility are necessary to enable exploration beyond the EVA walking range of the crew. An unpressurized mobility platform or platforms (such as the notional concepts shown in Figure 3-8) may contribute to meeting these needs, whether crewed or uncrewed. Functions that could be grouped into this capability category include providing local unpressurized crew surface mobility, as well as autonomous and/or tele- operations, and enabling additional science and utilization.

3.2.4.2 Pressurized Mobility

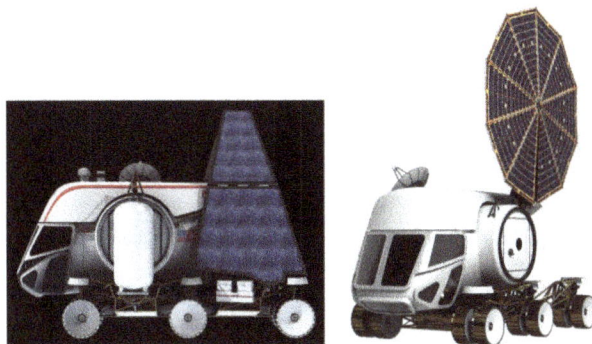

Figure 3-9. Example Concepts for Pressurized Mobility

Expanding surface mission durations and exploration range necessitates new capabilities in crew mobility and habitation. Combining both needs into a pressurized mobility platform or platforms is one option (as seen with the notional concepts in Figure 3-9), potentially encompassing functions such as crew habitation on the lunar surface, crew surface EVAs, pressurized crew surface mobility, crew IVA workspace on the lunar surface, logistics transfer (including fluids and gasses) on the lunar surface, autonomous and/or tele-operations, and mobile crew habitation. Additional functions may include enabling additional science and utilization, such as surface sample (including frozen samples) recovery, curation, and packaging. These surface mobility functions can be accomplished in many ways, e.g., mapping to a single element or to multiple elements, each covering a subset of the functions.

3.2.4.3 Cargo Transportation

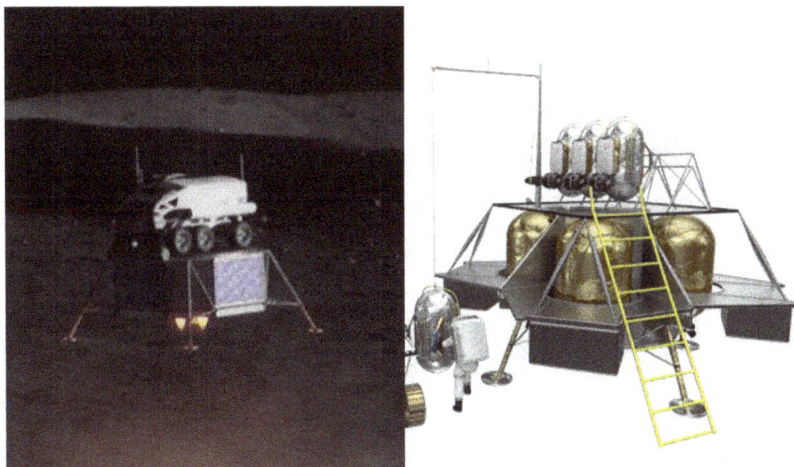

Figure 3-10. Example Concepts of Cargo Transportation

With periodic crewed missions and their expansion in duration and capability in addition to robotic presence, reliable transport of cargo to support the mission objectives is necessary. Some of the functions that may be grouped into this type of platform (with notional concepts shown in Figure 3-10) include the following: spacecraft aggregation in cislunar space, cargo delivery (e.g., science, utilization, technology, crew logistics) from Earth and unloading on the lunar surface, logistics transfer (including fluids and gasses), cargo return (including frozen samples) from the lunar surface to Earth, and in-space and/or surface cryogenic storage of propellant. Although platforms for delivering cargo were introduced in HLR, a growth in payload capacity to allow for delivering larger platforms like unpressurized or pressurized mobility assets or surface habitation assets is a potential expansion for FE.

3.2.4.4 Surface Habitation

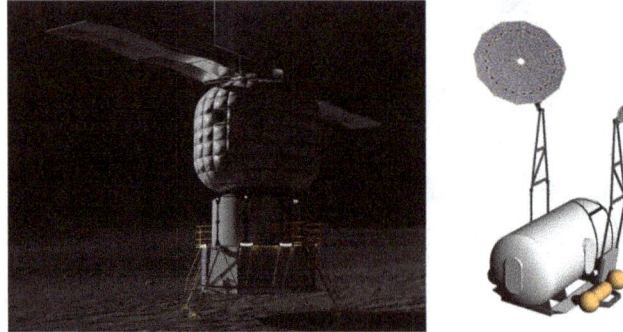

Figure 3-11. Example Concepts for Surface Habitation

With objectives related to long-term surface exploration, additional capabilities for crew surface habitation allow progress toward meeting those objectives. General crew habitation functions may be common across surface habitation platforms, like providing crew remote medical systems on the lunar surface, providing crew IVA workspace, enabling crew EVAs, providing long-duration deep space, partial-gravity crew habitation, and enabling logistics transfer, or these functions may be shared between several platforms of varying designs. Other unique functions that may be included could be ISRU production, storage, and/or transfer; demonstration of bioregenerative ECLS Systems; or demonstration of plant growth sub-systems. Some notional concepts are shown in Figure 3-11.

3.2.4.5 Orbital Exploration

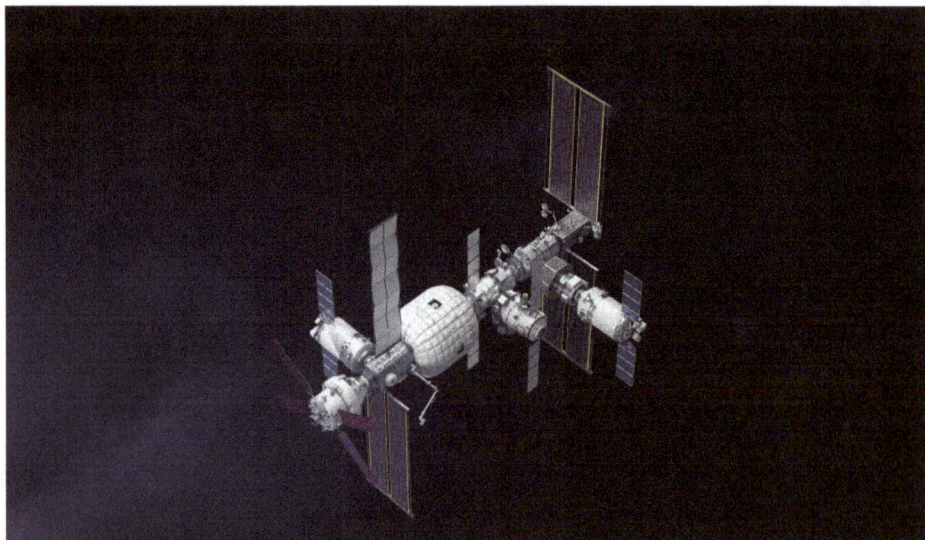

Figure 3-12. Gateway Expanded Capability Configuration with Visiting Expanded Habitation Example Concept

Foundational Exploration emphasizes extended duration and preparing for crewed Mars mission profiles through analog missions in lunar vicinity. An important aspect is long-duration mission segments in deep-space microgravity environment, which mimic the transit phases between Earth and Mars. To that end, a growth in cislunar orbital operations will occur as Gateway expands capabilities (as shown in Figure 3-12) and visiting vehicles, such as, a Mars transit habitat can be deployed. This expansion will support extended mission durations in preparation for Mars (e.g., objectives TH-3, TH-4, TH-8, HBS-1, HBS-2, HBS-3, OP-1, OP-4). Planned Gateway increased capabilities include the Gateway Extravehicular Robotic System (GERS) to be provided by the CSA, the ESPRIT Refueling Module to be provided by ESA, logistics resupply to be provided by JAXA, and a crew and science airlock module.

Sample functions that can aid in these areas include providing long-duration deep space microgravity crew habitation, crew remote medical systems in cislunar space, and crew IVA workspace in addition to enabling crew EVAs, crew vehicle docking, and spacecraft aggregation in cislunar space.

3.2.5 Sub-Architecture Descriptions

Forward work remains to further define the sub-architectures and their expansion for FE. In addition to integrating with particular elements, the sub-architectures bridge across elements and crew operations, necessitating high levels of coordination across the overall exploration architecture. As examples, notional, non-comprehensive functions for the communication, positioning, navigation, and timing and power sub-architectures are included here.

3.2.5.1 CPNT Sub-Architecture Notional Functions:

- High bandwidth, high availability communications between the lunar surface and Earth
- Provide position, navigation, and timing on the lunar surface
- Robotic systems on the lunar surface controlled from Earth and/or cislunar space
- Data storage on the lunar surface
- Precision landing systems for the lunar surface

3.2.5.2 Power Sub-Architecture Notional Functions:

- Power generation on the lunar surface
- Power storage on the lunar surface
- Power distribution on the lunar surface

3.2.6 Unallocated Use Cases and Functions

As the Foundational Exploration segment matures and elements, sub-architectures, and their associated functions are defined, mapping of FE use cases and functions to elements will be completed and the descriptions will be updated in future versions of the ADD.

3.2.7 Open Questions, Ongoing Assessments, and Future Work

With forward work remaining to define the Foundational Exploration segment, there are open questions on the segment, from the architectural approach(es) to accomplish the objectives to specific element and sub-architecture planning and design. The open questions here are non-comprehensive examples of the types of areas that will be addressed in future work and are only notionally binned for FE. This section will be updated in future revisions of the ADD.

- What is an attainable balance in mission types and locations to address infrastructure buildup objectives and scientific exploration of diverse sites objectives?

- What is the most effective way to utilize lunar missions as preparation for crewed Mars exploration?

- With the expansion of mission types and durations, what are the options for logistics resupply, both delivery to cislunar space and the lunar surface, and transfer to the necessary location(s)?

- What waste management and element repurposing, recycling, or disposal approaches should be utilized for sustainable exploration?

- What assets should be available to support non-polar sorties?

- What benefits do various levels of in-situ resource utilization provide for the lunar surface activities?

- How can the architecture expansion in FE enable key science and technology needs (e.g., polar volatiles, ISRU, biological, environments)?

- What options are available to significantly enhance sample return and conditioned/cryogenic cargo from the lunar surface to Earth?

- What assets should be available to support sustained scientific activity in the south polar region? How should they be distributed, and what are the supporting infrastructure dependencies?

Even as Foundational Exploration expands on what was accomplished in HLR, the FE missions set the stage for Sustained Lunar Evolution and make progress toward Humans to Mars.

3.3 SUSTAINED LUNAR EVOLUTION SEGMENT

3.3.1 Summary of Objectives

In the Sustained Lunar Evolution (SLE) campaign segment, NASA aims to build, together with its partners, a future of economic opportunity, expanded utilization, including science, and greater participation on and around the Moon. The focus on the segment is the growth beyond the Foundational Exploration segment to accommodate objectives of increased global science capability, long duration/increased population, and the large-scale production of goods and services derived from lunar resources. This segment is an "open canvas," embracing new ideas, systems, and partners to grow to a true sustained lunar presence. The steps for obtaining use cases for the sustained lunar evolution segment will involve broad coordination. Given the maturity of this segment, there is insufficient depth to allocate functions at this time beyond the higher-level capabilities associated with the objectives. However, for context, notional examples of the future use case and the sub-architecture dependencies over time are discussed as a placeholder for the initial work that needs to be completed.

Sustained lunar presence represents responsible long-term exploration of the surface and the establishment of a robust lunar economy. This segment is driven by RT-9 (Commerce and Space Development: foster the expansion of the economic sphere beyond Earth orbit to support U.S. industry and innovation), TH-3 (Develop system(s) to allow crew to explore, operate, and live on the lunar surface and in lunar orbit with scalability to continuous presence conducting scientific and industrial utilization as well as Mars analog activities), and the infrastructure objectives with the overarching goal of "Create an interoperable global lunar utilization infrastructure where U.S.

industry and international partners can maintain continuous robotic and human presence on the lunar surface for a robust lunar economy without NASA as the sole user, while accomplishing science objectives and testing for Mars." A sustained architecture at the lunar surface would further enable achievement of key science objectives in Lunar/Planetary Science, Heliophysics, Human and Biological Science, and Physics and Physical Science, as well as facilitate addressing new science objectives identified as a result of discoveries made during the previous campaign segments.

3.3.2 Use Cases and Functions

Architecting from the right requires the development of use cases that are coordinated with NASA's partners and based in economic plausibility in order to derive the functional needs. Table 3-15 below is an example set of interconnected notional paths worked in parallel to incrementally achieve sustained states of increased duration and population, increased economic opportunity, and increased science capability as guided by the objectives and recurring tenets. Future work will involve developing uses cases in coordination with NASA's partners.

Table 3-15. Example Sub-Architectures and Use Case Evolution for SLE Segment

Foundational Exploration Segment	Sub-Architecture and Use Case Evolution			Notional SLE Use Cases
Foundational Capabilities for: • Lunar Surface Access • Mobility • Habitation • Logistics • Power • Manufacturing • Construction • In-Situ Resource Utilization / Production	Expanded Power for Expanded Missions *More mission opportunities further from the South Pole for longer durations*	Increased Crew Size & Duration *Replicated surface habitats, laboratories and increased logistics*	Permanent Lunar Outpost *Crew/cargo access to and from the lunar surface enabled by ISRU, scores of crew*	Increased Duration & Population
	Minimal ISRU & Regolith Utilization *100s of kg of water/propellant produced*	ISRU Derived Propellants *1,000s of kg of water/propellant produced, minor civil engineering*	*Industrial-Scale ISRU & Mining* *10,000s of kg of ISRU propellant with regolith used for raw materials, 3D printing, propellant manufacturing, and mining*	Increased Economic Opportunity
	Expanded Mobility & Range *10s of km to 100s of km range from South Pole*	Increased Sample Return *100s of kg from non-polar regions cached and returned to Earth in addition to FE capabilities*	Lunar Global Access (Crew & Cargo) *1,000s of kg from global locations returned to central location, then returned to Earth*	Increased Science Capability

3.3.2.1 Increased Science Capability

The science objectives are supported by the ability to deliver science instruments to various locations in cislunar space and the lunar surface and return the acquired data or samples to Earth. In addition, providing real-time human interaction where science activities are being performed increases the ability to rapidly react to discoveries and to determine optimal areas and samples to explore. When coupled with the ability to update, replace, and repair the systems for performing the science, human presence is extremely beneficial. Prior to this segment, science capability is governed by the initial orbital platforms, landers, and regional exploration infrastructure coupled with the HLS ability to support global lunar sorties, including to the lunar far side. Although the Foundational Exploration segment will have the function to return the required science samples gathered during a 30-day-class mission, approaches to increase the science capability as mission duration and available power beyond the previous segment's limits will have to be addressed. A notional path working across the sub-architectures to increase science abilities beyond the previous segment is discussed next, in the context of the objectives and key characteristics.

Increasing science capability is enabled by enhancing multiple sub-architectures, with trades within those architectures to understand the best approach. If global concurrent lunar science activities represent the desired end state, then the lunar communications and navigation sub-architecture will need to evolve via interoperability, scalability, and reconfigurability to allow concurrent science missions distributed across the lunar globe to send back data on high-speed links. This would represent a continued evolution beyond the initial communications/navigation infrastructure that features direct-to-Earth for the lunar near side, relay service for the South Pole region and limited relay services for non-South Pole regions. NASA and its partners can trade different approaches for satellite constellations, surface relay infrastructure and technologies such as optical links to enable high-data-rate communications.

Working backwards from and forward to the notional use cases across the segments informs key sub-architecture questions like what access and purity for viable ISRU are needed; what power interface and standards can enable a power grid that evolves to industrial scale; and what communications, navigation, and positioning architecture features will be required to scale to an evolved lunar future.

3.3.2.2 Increased Economic Opportunity

Economic opportunity on and around the Moon in the context of this discussion means that governments are no longer the sole source of support for the funding of the lunar activities and that non-governmental entities would like to invest in, and profit from, activities at the Moon. NASA aims to reduce the barriers of entry for activities on and around the Moon and to provide capabilities others can leverage. Currently there is limited economic rationale for exploring the Moon, but given the cost of getting to and from the Moon, knowledge and access are perhaps the first areas where economic opportunity exists for the non-governmental sector. Artemis is making the foundational investments for access to the Moon from a transportation, exploration, and science perspective. The opportunity for industry at this point is to leverage that investment to enable lunar access (both robotic and human) to additional governmental entities, scientific institutions, international entities, and industry partners. As mentioned in the increased science capability narrative, additional investments in communications, navigation, ISRU, power, and transportation sub-architectures will be needed to enhance access and return, facilitating the beginning of new supporting service economic opportunities in those areas.

Economic opportunity/profitability could progress along the lines of 1) information transfer, 2) delivering goods, 3) providing services at the Moon to enable others, and 4) bringing goods from the Moon to other destinations. Larger-scale economic opportunity begins to emerge when lunar reach and access are expanded, as small-scale ISRU propellant grows to industrial-scale, as aggregate power grows from kilowatts to megawatts, and using in-situ material and as manufacturing becomes more economical than importing everything from Earth. Once ISRU production is of sufficient scale, exporting propellant and material beyond the lunar surface is manifested as an economic opportunity.

3.3.2.3 Increased Duration and Population

Increased science capability influences economic opportunity, which overlaps with the need to increase both the population of humans at the lunar South pole region and the need for them to stay there longer. However, humans currently require a significant quantity of resources imported from Earth to survive, along with large amounts of pressurized volume in which to live safely. In order to increase the size and duration of the lunar population significantly, local resources will eventually be required to provide water, support food growth, and build out infrastructure, with commercial or internationally provided crew transportation systems infused to increase mission frequency and crew population. As an interim step, small modular systems could be supplied by multiple partners to act as a bridge between the initial Foundation Exploration capabilities and the full-up ISRU systems to provide additional habitation and logistics. Fission power augmentation will also be required to achieve a year-round population at the lunar South Pole region, as available sunlight oscillates as a function of month and season. At some point in this evolution the possibility of lunar tourism appears, possibly at first with Earth-provided modular systems at a higher cost, then later at a larger, more affordable scale once lunar resources can be fully leveraged.

3.3.3 Reference Missions and Concepts of Operation

Given the maturity of this segment, it is future work to define reference missions and detailed concepts of operation and will be defined as the architecture matures.

3.3.4 Elements and Sub-Architectures

Although the notional use cases discuss the implications of sub-architecture evolution across those use cases and time, actual element functional allocation and sub-architecture evolution will require the development of the use cases by the appropriate stakeholders before further decomposition can be performed.

3.3.5 Open Questions, Ongoing Assessments, and Future Work

Increased science capability, economic opportunity, and duration/population at the lunar South Pole region have the potential to evolve and merge in the future to form the first sustained human civilization beyond Earth. The capabilities put in place during the initial Artemis segments feed forward and enable the future enhancements, and the partnerships forged grow to incorporate a broader community. As Artemis solidifies its implementation of the previous segments, planning for the SLE segment needs to begin in earnest, as the ideation of both the future lunar state and the path(s) for getting there will impact what comes before it. Given the objective decomposition process as described in Section 1.3.1, the notional use cases and functions described in this section need to be replaced with ones developed by the segment stake holders.

3.4 HUMANS TO MARS SEGMENT

3.4.1 Summary of Objectives

The Agency's M2M Strategy laid out specific tenets and goals to guide the development of an integrated Moon-to-Mars architecture. In addition to the cross-cutting science and operations goals, Mars-specific goals in both Infrastructure and Transportation & Habitation provide further architecture implementation guidance.

3.4.2 Use Cases and Functions

The decomposition of objectives into characteristics and needs and then to use cases and functions for the Mars segment has not been fully completed. High level functions needed for each of the systems are described in the subsequent sections, serving as a starting point to lay the foundation for a Mars architecture decision roadmap.

3.4.3 Mars Trade Space, Reference Missions, and Concept of Operation

For the purpose of initial analysis, a technically feasible early practical Mars mission was used. For example, a Mars surface mission would be too challenging for a solo explorer, so two crew to the surface is the current practical minimum working assumption. However, the trade space remains wide open. Definition of the full trade space, along with updated reference missions and concept of operations, are being developed in the context of the Moon-to-Mars Objectives and will aid in developing the Mars architecture decision roadmap.

3.4.3.1 "How" to Get to Mars and Back?

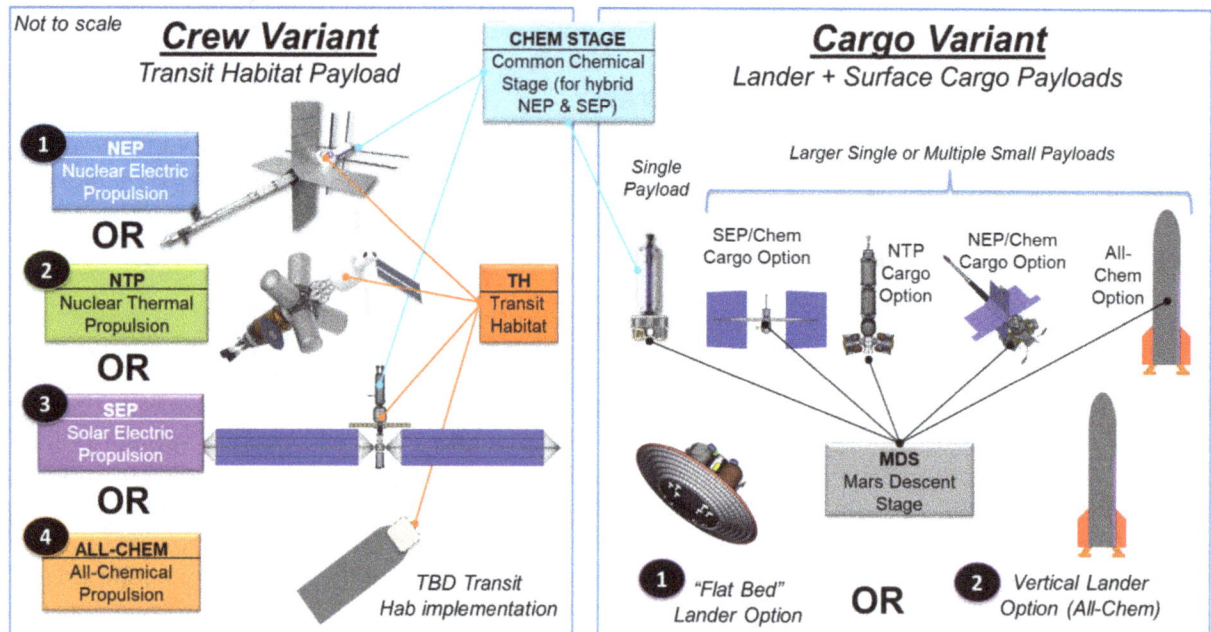

Figure 3-13. Major Mars Architecture Transportation Options Trade Space

Because the first challenge of any Mars mission is simply to get to Mars safely and return to Earth, the Earth-Mars transportation system elicits a substantial amount of discussion relative to the "How?" trade space. To that end, recent analysis was designed to explore the pros and cons of different transportation system options across a wider range of mission profiles than previously considered. The initial metric of interest for recent assessments was total roundtrip mission duration, due to the significant duration-related flow-down impacts to crew health and performance, technology investment, development timelines, and cost. Historically, Mars mission duration has been treated as a binary choice: either an ~2-year-long opposition-class mission characterized by at least one high-energy transit leg and a very short Mars stay measured in days, or a 3-or-more year conjunction-class mission characterized by low-energy transits with at least a year-long loiter period at Mars. In truth, mission duration may be thought of as a continuum: the architecture can be optimized for any given duration for a particular opportunity year or a range of durations over different opportunities.

To inform the total mission duration decision, which in turn will inform a host of other decisions (including transportation propulsion technology investments), stakeholders will need several pieces of information: an understanding of system-by-system performance sensitivity over the entire duration trade space and an integrated campaign and risk assessment for the various possible implementations, including integrated risks to the human system. To that end, the Mars architecture concepts presented here are intended to populate a broad swath of the "how" trade space, allowing decision makers to see how different implementations of four different transportation systems fare in the context of different reference missions (the "Why," "What," and "Where").

As shown in Figure 3-13, transportation system concepts currently under evaluation include hybrid Nuclear Electric Propulsion/Chemical (NEP/Chem), Nuclear Thermal Propulsion (NTP), hybrid Solar Electric Propulsion/Chemical (SEP/Chem), and All-Chemical (All-Chem). For comparison purposes, a common transit habitat (TH) is assumed for all crew transportation systems in the crewed variant; cargo variants of each concept are also available. Two different Mars Descent System (MDS) concepts have also been developed: a relatively small 25 metric ton (t) payload capacity "flat-bed" lander and a larger vertical lander capable of landing the minimum total surface payload cumulative mass of 75 t. For comparison purposes, a common set of surface systems is assumed for all architectures, as is a common Mars Ascent Vehicle (MAV) concept. To bound the trade space, recent analysis has focused on a minimal 2-crew MAV concept that relies on Earth-delivered ascent propellant, but more complex options capable of ferrying larger crew complements using ISRU propellants have been studied and will be revisited for later sustained exploration missions. Details of these concepts are provided in subsequent sections of this document.

3.4.3.2 Mars Initial Analysis Assumptions

Human Mars mission requirements will be developed under an eventual human Mars exploration program. In lieu of requirements, guidance provided by NASA leadership heavily influenced the architecture concepts used for human Mars mission architecture development.

Recent analysis assumptions used to assess impacts for the Mars exploration campaign's architecture development include the following:

- A light initial exploration footprint: four crew members to Mars orbit with two crew members descending and living on the surface for a 30-sol surface stay

- Multiple Mars landers, with the first lander(s) pre-deploying cargo to prepare for a later crew landing

- Modest initial surface infrastructure: a 10 kWe minimum FSP system and communications infrastructure, but no surface habitat, and no return-mission-critical ISRU propellant production

- "All-up mission" approach: crew depart Earth with all the transit propellant they need for the round-trip journey, a consequence if there is no-ISRU for early missions

Note that these assumptions are considered for a basis of comparison only. More complex mission scenarios will be addressed in subsequent analysis cycles, but the initial step is to define a practical architecture for the first human Mars mission campaign from which to expand. It is important to note that none of these assumptions are fixed; they provide a framework for direct architecture comparisons, and all decisions will be made with architecture evolution in mind.

3.4.3.3 Reference Missions for Assessments

To provide stakeholders with a sense for how the Mars architecture changes as just a single constraint is varied, three reference missions of different total durations— but all with the same surface and transit operational constraints, such as environmental exposure, communication delays and blackout periods —are defined to enable assessment of the architecture to inform the eventual decision roadmap (Figure 3-14): Reference Mission 0 with an Earth-Mars-Earth transit duration not to exceed 760 days, Reference Mission 1 with a moderate transit duration of 850 days, and Reference Mission 2 with a more relaxed transit duration of up to 1,100 days.

Reference Mission 0 is an opposition-class type of mission where at least one leg of the transit requires substantial energy to close the distance gap between Earth and Mars rather than loitering in Mars orbit to take advantage of planetary motion as in Reference Mission 2. Reference Mission 0 reflects the desire to shorten the roundtrip mission duration in an attempt to reduce long-duration spaceflight risk to the crew.

Reference Mission 2 represents the traditional conjunction-class corner of the trade space, taking advantage of minimum-energy trajectories by loitering in Mars' vicinity for up to a year, which in turn reduces overall propellant mass and launch costs. This reference mission represents the desire to minimize the total mass of the transportation system by minimizing the energy required for the roundtrip journey.

Mission Duration Knob	WHO We Send	WHAT We Do	WHERE We Go	WHEN We Go	WHY We Go	HOW We Get There & Back
Fast Roundtrip High Energy **Reference Mission 0**	**Analysis Assumption:** **Number of Crew** **Range of 2 – 6**	**Analysis Assumption:** **75t Total Landed Payload** Light footprint: Minimal surface infrastructure, crew live in rover, 10 kWe Fission Surface Power (no return propellant ISRU)	**Analysis Assumption:** **Single Mars Surface Site +35° N Latitude**	**Analysis Assumption:** **2039 crew departure to meet "by 2040" boots on Mars**	**Science** **Inspiration** **National Posture**	**Mission Time:** 760d or less in Deep Space, fixed 50 sols in Mars Orbit, w/30 sols on Mars Surface (870-900 days total crew time off Earth)
Moderate Duration Moderate Energy **Reference Mission 1**						**Mission Time:** 850d in Deep Space, fixed 50 sols in Mars Orbit, w/30 sols on Mars Surface (960-1020 days total crew time off Earth)
Long Duration Minimum Energy **Reference Mission 2**		**>75t Total Landed Payload** Light Footprint: Plus leverage additional capacity if available				**Mission Time:** 950-1100d in Deep Space, no less than 50 sols in Mars Orbit, w/30 sols on Mars Surface (1090-1250 days total crew time off Earth)

Figure 3-14. SAC22 Humans to Mars Reference Missions for Transportation System Assessments

Reference Mission 1, though accelerated, is not strictly an opposition-class mission; rather, it is on the continuum between traditional opposition-class and conjunction-class missions. This reference mission represents a compromise between Reference Mission 0 and Reference Mission 2, in an attempt to understand the middle ground of this particular trade space.

The rationale for multiple reference missions is twofold: first, to assess candidate transportation propulsion system performance over the continuum from opposition-class to conjunction-class missions, and second, to answer the question "Is nuclear propulsion needed to enable crewed Mars missions?" A given propulsion concept may perform well for one mission class type, but not others; by considering different mission class types, decision makers can better compare these architectures to each other and understand how constraints such as total mission duration influence performance.

3.4.3.4   Mars Architecture Decision Categories

Surface Systems	EDLA Systems	Transportation	Crew Support
Major Mars Architecture Decisions (full list in work)			
Surface Mission Purpose	**Payload Mass:** descent/ascent	**Duration:** total crew time away and Mars orbit time	**Number of Crew**
Number of Crew	**Largest Indivisible Payload Mass and Volume**	**Mission Mode**	**Communications Strategy**
Stay Duration		**Mars Parking Orbit**	**EVA Strategy**
Habitation Strategy (linked to duration/crew)	**Orbital Crew Ops:** Split crew vs. all to surface	**Number of Crew**	**Logistics Strategy:** including disposal ops
Crew Ingress/ Egress Strategy: includes planetary protection, logistics	**Landing Site:** one vs. multi-site, no. of missions	**Habitation:** Integrated vs. Separable, no. of crew	**Crew Systems Strategy:** including health care, hygiene, food system, etc.
Mobility Strategy: exploration radius	**Return Propellant Strategy:** storable vs. cryogenic propellant and Earth-delivered vs. Mars ISRU	**Propulsion Tech:** Nuclear-Enabled vs. Non-Nuclear, linked to duration and need date	
Communications Strategy		**Stack Aggregation Strategy/Location**	
Surface Power: sizing	**Element Re-use**		
Element Re-use: linked to no. of missions/site selection	**Number of Crew**	**Element Re-use:** linked to mission cadence	

Figure 3-15. Sample of the Major Mars Architecture Categories and Decisions

To aid in assessing extensibility of Mars elements to other destinations or programs and vice versa, the Mars architecture elements can be bucketed into four major categories: 1) Mars surface systems that enable crew to live and work on the planetary surface; 2) Entry, Descent, Landing, and Ascent (EDLA) systems that are able to move crew and surface systems from Mars orbit to the Mars surface, and return crew and cargo back to Mars orbit; 3) transportation systems that are able to move crew and cargo from Earth to Mars orbit and back again; and 4) crew support systems that cross multiple missions, phases, and destinations, such as EVA spacesuits, distributed communications networks, or crew healthcare systems. These categories, and the key architecture decisions required within each category, are outlined in Figure 3-15.

As noted above, major decisions (the "Why" or "When," for example) will heavily influence subsequent decisions within each architecture category. Because decisions in one architecture category will ripple across the other categories as mass, cost, or complexity, NASA will study the effects of options across the end-to-end architecture, under various decision structures. This process will enable NASA to develop a roadmap of key architecture decisions. It is important to note that Figure 3-15 and the description provided in this document are not intended to provide an exhaustive list of decisions and categories, but rather to begin development of the integrated Mars architecture decision roadmap for eventual implementation. Key decisions that will affect all 4 Mars architecture elements are establishment and approval of the Agency loss of crew (LOC) Safety Reporting Thresholds (SRTs) and directorate loss of mission (LOM) requirements. The LOC SRTs and LOM requirements specify the minimum tolerable/allowable levels of crew safety (maximum tolerable level of risk) and mission loss, respectively, for the design in the context of the proposed design reference mission(s). These are key early steps in the human-rating certification process that will aid in allocating reliability requirements and identifying safety/risk technological areas that require further development, prioritization, and/or demonstration.

3.4.4 Mars Surface Systems

The initial focus for Mars exploration is the development of a modest first exploration mission, framed as a first step to a sustained human exploration campaign. For the sake of apples-to-apples comparisons, initial analysis assumes the same surface system elements regardless of how those systems are transported or deployed to the Martian surface. Long-duration crew stays at Mars will be assessed as future work related to sustained human exploration analysis.

3.4.4.1 Functions

The primary function of human Mars surface systems is to protect crew and utilization payloads from the Mars environment during the Mars surface mission duration. Mars surface systems will also be critical to enabling science investigations before the crew arrives, while they are on the surface, and after they depart. For utilization payloads, this includes the pre-deployed cargo phase prior to crew arrival, as well as an extended robotic operations phase following crew departure. Capabilities required to perform this function include utilities such as power and communications, surface mobility assets, and habitable volumes.

3.4.4.2 Key Decisions and Drivers

Virtually every other surface system decision will hinge on the desired number of crew members on the surface ("Who?") and their purpose ("Why?"). That is not to say that other decisions could not be made first, but these two decisions may be considered anchoring decisions for a logical flow of subsequent surface architecture decisions. Understanding the interrelationship between these decisions is vital in developing an integrated surface architecture.

3.4.4.2.1 Surface Mission Purpose

The Mars surface architecture will vary significantly depending on whether the surface mission purpose is confined to a narrow set of specific human-assisted science objectives, a set of tasks intended to lay the groundwork for sustained human presence, or something else. Key considerations will be high-priority science objectives, technology demonstrations, and long-term exploration plans.

Note that decisions involving the surface mission purpose will have impacts beyond the surface architecture. For example, surface mission purpose will inform landing site selection, which will drive the EDLA, and transportation architectures. Also note that science, technology demonstration objectives, and utilization strategy is, by definition, a key factor in the surface mission purpose decision, with flow-down impacts to landed utilization mass, volume, and power; this factor potentially influences the payload capacity and/or number of landers required.

3.4.4.2.2 Number of Crew Members to the Surface

The minimum practical number of crew members to be sent to the surface is assumed to be two, given NASA's long-standing "buddy rule" for critical spaceflight operations. Initial crew complements as high as six have been analyzed, but ultimately the number of crew members required will be tied to the surface mission purpose, with more crew members needed for more elaborate mission plans. Whether to split crew (with some remaining in orbit while others descend to the surface) will depend on orbital and surface tasks and may require iterative analyses to assess different architectures and concepts of operation.

3.4.4.2.3 Surface Stay Duration

Minimum surface stay duration will be a function of the surface mission purpose and how many crew are available to accomplish the mission. Depending on the architecture, there may be logical stay duration break points, beyond which an additional element may be required to complete the mission.

3.4.4.2.4 Habitation Options

There are multiple habitation options to consider that are largely based on architecture decisions regarding crew size, interfacing surface mobility strategy, EDLA and transportation capability, and the location of initial and subsequent crewed Mars missions. Numerous studies have analyzed the necessary habitable volumes for a surface crew, for various mission scenarios, as well as the EDLA and transportation architectures that must deliver these elements. Habitat re-use or re-purpose from previous missions must also be considered as part of the surface habitation strategy.

3.4.4.2.5 Crew and Logistics Ingress/Egress

Habitation decisions will in turn inform crew ingress/egress options, with key considerations being dust mitigation, planetary protection, logistics management, contingency access, system maintenance needs, and operational efficiency. Crew and logistics ingress/egress strategy is expected to be heavily informed by Artemis experience on and around the Moon, along with mission-specific constraints such as schedule, mass, and cost. Options explored include ingress/egress via airlocks, hatches directly into the habitable volume, and/or use of a (suit) port that allows crew to directly don a spacesuit via a detachable hatch mounted directly to the exterior of the habitable volume.

3.4.4.2.6 Surface Mobility Options

Surface mobility decisions will be derived from the mission purpose (where do we need to go to meet the objectives and what do we need to do there?), stay duration (how long do we have to get there and back to the MAV?), cargo movement (what payload elements need to be moved from one location [e.g., the lander deck] to other locations?), and habitation decisions (are the habitable volumes moveable? what are the traverse distances to/from habitat, landing site, and ascent stage?). Each of these individual considerations will influence the overall exploration radius. Because mobility includes EVA systems, crew ingress/egress must also be considered as it will influence EVA suit design and operation. Mobility systems may also be required to support autonomous or remotely commanded operations before the crew arrives or after they depart, which may influence the communications architecture. Vehicle mobility systems, such as a pressurized rover, tend to be large, so mobility decisions will impact the EDLA and transportation architectures that must deliver these elements.

3.4.4.2.7 Surface Communication Options

Surface communications decisions will ultimately depend on how many surface assets are deployed; their relative proximity to each other and whether there are potential line-of-sight obstructions between them; the mobilization plan as assets move around the surface; which assets need to communicate with each other, with orbiting assets, and/or with Earth; data rates required between various assets; and power available to each asset. Surface communications decisions are expected to be heavily informed by Artemis experience on and around the Moon, and there is likely to be iteration across the elements as mass, power, complexity, or other constraints are balanced.

3.4.4.2.8 Surface Power Options

Fission Surface Power is the leading candidate for primary Mars surface power, due to the prevalence of dust storms that have proven difficult for solar-powered Mars surface systems. Although solar-powered short-duration surface missions might have acceptable risk, longer-stay missions, or missions pre-deploying powered cargo will likely require surface power technologies, such as FSP, that are resistant to environmental disruption. Initial analysis has identified a minimum power level needed to achieve a 30-sol, light footprint exploration mission, but it remains forward work to assess and integrate power needs for science and technology demonstration. Therefore, the power level, number of units, and operations plan (leave surface power system on the lander it arrives on or deploy elsewhere) remain as forward work.

3.4.4.2.9 Surface Architecture Life and Reuse

Operating life limits, including re-use for subsequent missions, will depend on total surface mission duration (including the pre-deployed cargo and post-crew departure robotic science mission phases) and whether subsequent human missions will return to the first mission landing site.

3.4.4.2.10 Return Propellant Strategy

Return propellant strategy—whether manufacturing propellant in-situ on Mars derived or Earth delivered resources, or using propellant delivered from Earth and potentially moved around the surface to fuel different vehicles—drives surface system mass, power, operational timelines, and potentially landing site selection.

3.4.4.3 System Concepts

Table 3-16 summarizes the minimum set of surface concepts based on functional needs currently being evaluated in the Mars surface architecture. Note that these are at the conceptual design phase; subsequent publications will contain additional surface concept reference detail as they mature, and more concepts may be added as additional functions are defined.

Table 3-16. Mars Surface System Functions and Example Concepts

Functions	Example Concept(s)		Heritage and Status
Provide power for all surface elements	Surface Power		FSP derivative of the Kilopower concept is in formulation for a lunar demonstration mission. Other options have been evaluated but may not meet constraints.
Provide power storage and distribution from the source to surface end-users or distribution points			Derivative of cables used in Earth applications (e.g., solar farms, off-shore wind farms, under sea cabling). Various deployment concepts are being evaluated.
Enable automated cargo handling	Robotics		Robotic cargo handler: simple heritage design and components. Should be scalable to multiple cargo types or sizes
Enable autonomous or remoted-controlled fine motor-control manipulation of mechanisms and other components, such as cables, hoses, etc.			Robotic manipulator heritage from ISS robotics, Robonaut, etc.
Provide aerial exploration and contingency support			Advanced generation of Ingenuity, a robotic helicopter landed with Perseverance and currently in use on Mars
Provide ability for crew Extravehicular Activity	Refer to Mars Crew Support Architecture, Section 3.4.7		
Provide habitable volume for shirt-sleeve crew for surface activities	Habitation		Lunar habitation and mobility derivative concept

Functions	Example Concept(s)	Heritage and Status
Provide ability to transport crew utilization payloads, and other cargo across the Mars surface. Note: Utilization payloads can include science equipment.	Surface Mobility	Mars terrain vehicle derivatives of lunar mobility concepts
Provide ability to store, condition, and transfer ascent propellant through Earth launch, transit, and Mars landing, as well as the Mars surface environment	Propellant Storage and Transfer	Same design and fluid-compatibility as conceptual ascent vehicle propellant tanks
Provide means to transfer propellant to the ascent vehicle	Propellant Delivery	Potential concept leverages technology developed in support of the satellite servicing mission previously called Restore-L (now OSAM-1) and its predecessor Restore-G
Science: Provide equipment needed to meet Mars surface-based science objectives		To be coordinated with the Science Mission Directorate
Technology Demonstration: Provide equipment needed to meet Mars surface infrastructure technology demonstration objectives		To be coordinated with the Space Technology Mission Directorate
Return Cargo: Provide containment and environmental control for Mars-origin or Mars-contaminated materials		To be determined based on mission objectives
Communication between crew, surface assets, orbital assets, and Earth	Refer to Mars Crew Support Architecture, Section 3.4.7	
Provide logistics	Refer to Mars Crew Support Architecture, Section 3.4.7	

3.4.4.4 Concept of Operations

HEOMD-415, *Reference Surface Activities for Crewed Mars Mission Systems and Utilization*[14], provides detailed information on initial surface mission concepts of operation. As noted above, no decisions have been made, and significant forward work remains to define science and technology demonstration objective implementation options and integrate with crew systems and operations.

---

[14] *Reference Surface Activities for Crewed Mars Mission Systems and Utilizations,* National Aeronautics and Space Administration (2022). HEOMD-415.

3.4.5 Mars Entry, Descent, Landing, and Ascent Systems

All surface system assets, plus the crew's ascent system, must be descended and landed on Mars. The landed mass required for a human mission exceeds the practical limits of heritage robotic mission EDL systems such as parachutes, airbags, or sky cranes. Two different types of landing systems are currently being assessed: a "flat bed" lander where the payload is mounted on a cargo deck relatively close to the surface and a "vertical lander" that could accommodate higher-mass payloads. The number of landers needed for a particular mission will depend on the lander's payload capacity (both mass and volume) and any pre-deployment timing constraints. For the purpose of apples-to-apples comparison, it is assumed that both types of landers could deliver the same surface cargo, including the same surface and ascent system with Earth-origin propellants. There are alternative ascent system schemes employing in-situ propellant manufacturing, but because these options stray from the first mission's "light exploration footprint" assumption, those options are deferred to subsequent analysis cycles.

3.4.5.1 Functions

Regardless of design, all Mars EDLA systems must provide a minimum set of functional capabilities to support the integrated Mars architecture.

3.4.5.1.1 Protect Crew and Cargo During Mars Entry, Descent, and Landing

The Mars EDL system must accommodate rapid changes in temperature, pressure, and gravity while decelerating from orbital velocities without transmitting damaging loads to crew or cargo. Due to a combination of potential crew deconditioning, lengthy communications delays with Earth, and the rapid pace of dynamic events during EDL, Mars EDL systems must be designed for autonomous operation with limited real-time crew input.

3.4.5.1.2 Protect Crew and Cargo During Mars Ascent

The Mars ascent system must accommodate rapid changes in temperature, pressure, and gravity without transmitting damaging loads to crew or cargo. The ascent vehicle will also be responsible for providing a habitable environment to support the crew during ascent from the Martian surface.

3.4.5.1.3 Protect Against Cross-Contamination of Martian and Earth Environments

Descent systems will need to minimize the transfer of uncontained Earth material to prevent forward-contaminating the Martian environment to maintain pristine scientific samples to the maximum extent possible. Similarly, ascent systems will need to minimize the transfer of uncontained Martian material to prevent backward-contaminating Earth return vehicles.

3.4.5.1.4 Provide Integration Interfaces to Mars Transportation and Surface Systems

Mars EDLA elements must receive services (such as power, data, or thermal control) from the Mars transportation system during transit and must provide similar services to the cargo payloads they carry during transit, entry, descent, landing, and surface operations prior to accessing power from surface infrastructure (i.e., from the FSP).

3.4.5.1.5 Provide Precision Landing Capability to Enable Multi-Lander Surface Operations

Lunar landing systems will be insufficient to meet precision landing requirements on Mars. Technologies developed for the Moon may be applicable but insufficient due to differences in EDL on Mars, primarily due to the presence of the Martian atmosphere. Architectural decisions such as Mars parking orbit can also impact the required precision landing technology development due to different flight path angles and relative velocities during key phases of EDL. Additionally, there may be constraints on landing precision technology imposed by the need to land multiple landers in close proximity for operational purposes while simultaneously maintaining safe distances from previously landed assets to mitigate the potential damage caused by ejecta lofted when terminal descent rocket engines interact with surface materials.

3.4.5.2 Key Decisions and Drivers

EDLA design will be heavily influenced by two human Mars mission requirements and two constraints. Total required payload mass to the surface and back to Mars orbit is informed by "Why" we are going to Mars and "What" we will do there, which in turn drives the number of crew members and equipment we need to land, as well as the number of crew members and cargo we need to return to orbit. Whether these systems need to be extensible to larger future payloads may also influence EDLA design. EDLA design will also be constrained by the largest indivisible payload item (mass and volume) and whether the EDLA system is required to support split crew operations, where some crew members land while others remain in orbit.

3.4.5.2.1 Payload Mass Landed on the Mars Surface

The largest payload landed to date on Mars is about one metric ton, but even the most modest human Mars mission is estimated to require at least 75 t of total landed payload for even a short-duration surface stay. Longer, more ambitious missions will require more landed mass. Total landed payload mass, in combination with EDL technology availability, will determine how many landers are needed to complete the mission, which in turn will inform lander production, launch, and delivery cadence, with flow-down impacts to the Earth-Mars transportation architecture. The number of landers will also inform surface system concepts of operations, depending on how far apart landers are deployed, which payloads need to move between landers, and the power and communications strategy between them.

3.4.5.2.2 Payload Mass Ascended to Mars Orbit

Ascent from the Mars surface has never been attempted. The Mars Science Return Program Mars Ascent System is the first planned ascent from another planet. Mars atmosphere and gravity make this a high "gear ratio" operation, meaning several kilograms of ascent propulsion mass are required for every kilogram lofted back to orbit. At a minimum, ascending just two crew members—even without any return cargo—is estimated to require more than 30 t of propellant to a 5-sol Earth transportation vehicle parking orbit. Each additional kilogram of cargo mass further increases ascent propellant mass; either this mass must be added to the landed payload allocation noted above or propellant production mass and additional power must be added to the landed payload mass, with flow-down impact to the surface operations timeline.

3.4.5.2.3 Largest Indivisible Payload

Total landed payload mass can be distributed across smaller landers, which could minimize the EDL technology development burden, but the limiting factor will be the largest indivisible payload. For modest missions, this is likely to be the MAV. Propellant can be off-loaded onto other landers or manufactured on the surface to reduce landed mass. MAV hardware (tanks, engines, etc.) assembly is possible, but extremely risky, especially for initial missions. For more ambitious, longer-duration missions, a large surface habitat might be the pacing payload mass item, depending on how much/how fast outfitting could be installed after landing. Note that payload volume will be constrained by the payload shroud of Earth launch systems, potentially requiring additional Earth launches or Mars landers for physically large items that can be modularized, or larger launch and lander vehicles for those items that cannot be segmented and exceed current payload shroud size.

3.4.5.2.4 Orbital Crew Operations

If all crew members are to land on the surface, then direct entry options are possible, but if the architecture is required to support "split crew" operations (where some crew members remain in Mars orbit), then both the transportation and EDL systems may need to support orbital operations. The parking orbit has a significant impact on vehicle design, orbital operations, and timelines. Landing and ascent durations and time required to accommodate multiple launch/landing opportunities are highly dependent on the parking orbit. Additionally, mass of the MAV, which is already identified as a "high gear ratio" element impacting the design of several other architecture elements, is highly sensitive to parking orbit altitude. In general, EDLA systems favor lower parking orbits. However, in-space transportation systems tend to favor high parking orbits. Therefore, the optimal parking orbit is an integrated problem between EDLA systems, in-space transportation systems, and crew operations.

3.4.5.2.5 Landing Site Selection

The terrain of a selected Mars landing site location will obviously influence EDLA design, with landing site latitude and elevation affecting both ascent and descent propellant mass, with flow-down impacts to landed payload mass and surface operations related to MAV fueling strategy. Terrain and whether subsequent missions will return to a given landing site can also influence landing precision requirements. Key reconnaissance parameters (e.g. high-resolution imaging or surface properties assessments) may be needed to inform EDLA design. In addition, the landing site's lighting constraints during the descent phase of the mission could have integrated impact to the in-space transportation system.

3.4.5.2.6 Ascent Propellant Acquisition Strategy

Options include landing a fully fueled MAV on Mars or landing an empty or partially fueled MAV on Mars and either transferring propellant from another lander or manufacturing propellant from in-situ resources. All of these options result in flow-down impacts to other systems: a fully fueled MAV drives MDS payload capacity and Earth launch capacity; a partially fueled MAV drives surface propellant transfer mass and complexity; and in-situ propellant manufacturing drives surface system mass, power, and operational timelines. Constraints such as Earth launch fairing diameter and Mars parking orbit can also have significant constraints on ascent vehicle design choices and propellant acquisition strategy. For example, cryogenic propellant-based MAV to a 5-sol orbit challenges the geometry of an 8.4 m diameter Earth launch system fairing due to low density propellants combined with increased propellant loads for higher parking orbits. Workarounds to the Earth launch shroud constraint in turn impact the transportation system by potentially driving it to a lower Mars orbit (at a higher propellant penalty) or require the addition of a "taxi" element to bridge the gap between how low the transportation system can dip into the Mars gravity well and how high the MAV can ascend on a lighter propellant load.

3.4.5.2.7 Element Reuse

Reuse cannot be an afterthought for EDLA systems. It must be integral to the design. Feasibility of reusing EDLA systems is highly coupled between system design and concept of operation. Initial reference designs are not practical for reuse, but with changes to design and operation, reuse could be enabled. Certain designs may be more "evolvable" for reusability than others. Operating life limits, including re-use for subsequent missions, will depend on total surface mission duration (including the pre-deployed cargo and post-crew departure robotic science mission phases), and whether subsequent human missions will return to the first mission landing site.

3.4.5.3　System Concepts

Table 3-17. Mars Entry, Descent, Landing, and Ascent Functions and Example Concepts

Function	Example Concept(s)		Heritage and Status
Delivers crew and surface cargo from Mars orbit to Mars surface Serves as a launch pad for Mars ascent operations	Mars Descent System		MDS conceptual design available for "flat bed" type lander. HIAD based on Low-Earth Orbit Flight Test of an Inflatable Decelerator. Forward work to complete reference design of vertical lander.
Shirt-sleeve environment for transfer of crew and equipment between the deep space transport, habitation element, and MDS crew cabin (which may be a surface cargo element)	Pressurized Mating Adapter		Very high-level reference conceptual design available
Delivers crew and returns crew and cargo to Mars orbit	Mars Ascent Vehicle		Reference conceptual designs available
Shirt-sleeve environment transfer of crew and equipment between a pressurized surface asset and the MAV cabin	Surface Pressurized Tunnel		Very high-level reference conceptual design available

3.4.5.4　Concept of Operations

Refer to HEOMD-415[15] for the initial surface concepts of operation for various mission and architecture implementations, including MAV fueling strategies; that document will be updated as the architecture evolves.

3.4.6　Earth-Mars Transportation Systems

Earth-Mars transportation systems serve to transport the crew, surface systems, and EDLA systems to Mars and return crew to Earth. All Earth-Mars transportation architectures will consist of a propulsion and power backbone paired with one or more payload elements. For the purpose of this document, this integrated transportation system stack is referred to as the deep space transport (DST). A single DST design could be used for both crew and cargo deliveries, but to

---

[15] *Reference Surface Activities for Crewed Mars Mission Systems and Utilization,* National Aeronautics and Space Administration (2022). HEOMD-415.

optimize for cost, development schedule, or other metrics of interest, variants may be mixed within a single campaign: for example, a slower, less-expensive, non-nuclear transport for pre-deployed cargo with a faster, higher-powered nuclear system for crew transport. In the crew variant DST, the payload is a crew habitation system and all the utilization payloads, logistics, supplies, and spares for the in-space portion of the mission, including contingency operations. For the purpose of current analyses, a common habitation system is assumed for all transportation architectures. In the cargo-variant DST, payloads include surface systems, surface utilization payloads, EDLA elements, or other support system payloads.

Selection of a human Mars transportation system will be a complex decision shaped by numerous factors, such as mission objectives (the "Why?" question), exploration partner contributions and commitments, programmatic, schedules, and integrated risk assessments. The four transportation architectures presented here represent the range of options currently being analyzed.

Specific implementation of the different transportation systems will depend on the reference mission of interest and a balance between the optimization of the system and the robustness to other mission parameters. For each reference mission, transportation systems can be optimized, from both a configuration and a performance perspective, for the specific requirements of that reference mission. But an optimized transportation implementation might come at the cost of compromising the extensibility and flexibility to other mission design parameters that may be of interest.

3.4.6.1 Functions

Regardless of propulsion type, all Earth-Mars transportation systems must provide a minimum set of functional capabilities.

3.4.6.1.1 Provide Sufficient Energy to Transport Crew and Cargo from Earth Vicinity to Mars Vicinity and Back Again

The planetary alignment between Earth and Mars constantly changes over a roughly 15- to 20-year synodic cycle, so the amount of energy needed to make the transit will vary depending on the mission opportunity. If the transport is designed for only the "easiest" opportunity, Mars missions may be possible only once per synodic cycle; if designed for the "hardest" opportunity, the transportation system will be robust for all mission opportunities, but will be over-powered for most opportunities and will likely require more upfront technology investment. If carrying all required propulsive energy from Earth, the transport design must ensure that energy remains available throughout what will be a long round-trip mission duration; if planning to acquire return energy at Mars or an interim destination, the transport design must accommodate refueling or resupply operations with additional systems. To bound energy requirements, current analyses assume all propellant required for the round trip is launched from Earth and carried roundtrip, without need to resupply. For the purpose of sizing the transportation concepts, a complement of four Mars crew is currently under evaluation. This is likely a minimum practical limit for the purposes of addressing risk and redundancy; however, larger crew complements would require larger habitats and more consumables, which in turn will increase transportation energy requirements.

3.4.6.1.2 Protect Crew and Cargo from the Deep Space Environment for Transit Duration

In addition to the temperature extremes and near-vacuum pressure common in low-Earth orbit, Mars transit will have an additional complication of increased galactic cosmic radiation and prolonged microgravity risks. To protect crew and cargo during the long transit duration, the transport and integrated habitation systems must be sized to accommodate logistics and consumables with limited resupply options and spare parts to address routine and contingency operations. Note that as more mass is added to protect crew and cargo, more energy will be required to transport crew and cargo to Mars and back.

3.4.6.2 Key Decisions and Drivers

3.4.6.2.1 Total Mission Duration

The in-space transportation architecture is dictated by the celestial mechanics of Earth, Mars, and the Sun. The total roundtrip mission duration for a Mars mission is the primary driver for any in-space transportation decisions. Longer mission durations (~3 years) typically require lower energy, as they can rely on the more favorable alignments between Earth and Mars to perform two optimal transfers between the planets. Shorter missions would require more energy to complete, as the in-space transportation system will need to complete the roundtrip mission while fighting against the natural orbital energy of the two planets. The energy required, and therefore the propulsion technology and total propellant mass, scales exponentially with mission duration, so the shorter missions are exponentially harder than the longer missions. The total mission duration decision also cannot be made solely on the basis of the in-space transportation system; factors such as crew health and performance as a function of total mission duration must also be considered. This decision has broad implications to crew systems, crew health, Mars orbit time, and Mars surface time, which in turn will influence the scope of utilization activities.

3.4.6.2.2 Mars Vicinity Stay Time

The decision on Mars vicinity stay time is driven by three factors: the Mars surface mission duration, the Mars orbital operation requirements, and the total roundtrip mission duration. The minimum surface stay duration will be the minimum duration that the in-space transportation system needs to remain in Mars orbit. However, additional time is required to prepare and transfer crew to the MDS, prior to the surface mission, as well as time for the MAV to rendezvous, dock, and transfer crew and return Mars cargo back to the transport after ascent from the surface. Current assumption is 10 Martian sols prior to descent and 10 sols following ascent, totaling 20 sols, but assessment of operational needs and constraints is required to guide the final Mars orbit stay time decision. Finally, the total roundtrip mission duration will also have significant impact on the orbit stay time. Primarily, the shorter mission durations will have a lower bound for the orbit stay time, as the interplanetary trajectory is more energy efficient with more total duration in deep space, rather than in Mars orbit. For longer duration missions, the need to await optimal planetary alignment for the return journey will likely mean there will be a significant Mars orbit stay time available.

3.4.6.2.3 Mission Operation Mode

Mission operation mode refers to how the end-to-end mission is conducted and has significant implications to all other Mars decisions. Current assumption for the mission mode is an "all-up" mode, where the crew transportation stack departs Earth with all the propellant and logistics required to support the roundtrip mission. This is assumed for crew risk mitigation considerations, as the crew does not need to rendezvous with any propellant or logistics assets post-Earth departure to return safely, potentially descoping the surface mission in the event of an anomaly. Shorter-duration missions with higher energy may necessitate the pre-deployment of propellant at Mars to reduce the overall size of the transportation system.

To support the surface missions, the current assumption is that the surface assets are pre-deployed to the surface to wait for the crew. Potential options exist to integrate the surface elements with the crew stack so that no rendezvous is required in Mars orbit to support surface missions.

3.4.6.2.4 Mars Parking Orbit

The selection of the parking orbit at Mars for staging and aggregation of the mission will be dependent on the architecture and mission mode decisions, as well as surface abort timing constraints. Current assumption for Mars parking orbit is a 5-sol orbit, with the perigee of the parking orbit directly above the landing site to support a direct landing. This high-altitude parking orbit is beneficial to the transportation system because it does not require the whole transportation stack to insert deep into Mars' gravity well, but it puts an additional burden on the MAV, as the energy and time required to reach 5-sol orbit is higher than for a lower parking orbit.

3.4.6.2.5 Mars Landing Site

Related to the selection of the parking orbit, the final selection of the landing site will have significant impact on the transportation system. This decision will be interlinked with the Mars EDLA system. Assuming the EDL system does not have its own cross-range capability, the transportation system needs to deliver the EDL system to the appropriate parking orbit for descent to the surface. This could mean the transportation system needs to perform additional orbital maneuvers to change the orbital parameter of the parking orbit to align for both descent and potentially ascent portions of the mission. This impact is particularly profound for the crew transportation system as the integrated end-to-end trajectory needs both to bridge between the Mars arrival and departure interplanetary directions, and to satisfy the potential parking orbit constraint due to the landing site selection.

3.4.6.2.6 Number of Crew Members

The total number of crew members required for Mars will have a significant impact on the design of the transit habitation systems, which has flow-down effect on the transportation systems. Conversely, selection and design of the propulsion system will also impact the decision on the number of crew members due to maintenance, repair, and operational variations between the different transportation system options.

3.4.6.2.7 Transit Habitation System

The primary decision on the transit habitation system is about the integration between the habitat and the in-space transportation system. The habitation system can be integrated as part of the in-space transportation system or can be designed as an independent system. Current assumption for the transit habitation system is an independent system which will first facilitate early long-duration Mars precursor missions and Artemis activity in conjunction with Gateway before serving as the habitation system for the Mars missions. Another decision on the habitation system is whether the habitat should be a monolithic unit or modular in nature.

3.4.6.2.8 Propulsion Technology

There are multiple options for the transportation propulsion system. The decision will be informed by a plethora of other decisions, including total mission duration, transit habitation strategy, mission mode operation, and others. Nuclear vs. non-nuclear propulsion systems and high-thrust ballistic systems vs. low-thrust systems vs. hybrid high-/low-thrust systems are just a few of the propulsion technology decisions that must be made. However, propulsion system performance alone may not be a sufficient discriminator. The target date for a first human Mars mission will establish the propulsion system delivery date, which in turn will constrain technology development timelines—and may eliminate technologies that cannot be developed within the timeframe or dictate a phased strategy wherein early missions rely on available propulsion technologies and more advanced technologies are phased in during later missions. These key decisions will also be informed by non-technical considerations, such as the broader strategy question of long-term exploration objectives or technology development partnering arrangements with other Agencies, industry, or international partners.

3.4.6.2.9 Aggregation Location & Strategy

The Artemis and Gateway program lends itself to aggregation in cislunar space. The decision of where to aggregate the in-space transportation system and the strategy associated with it will be highly dependent on several factors, such as the selection and design of the propulsion system, the availability of different launch vehicles and their associated launch mass and volume capability, and the launch cadence of the aggregation campaign. There are optimal aggregation location and strategies for each of the transportation system options based on the launch vehicle cadence and capability. The integrated nature of the aggregation strategy complicates the decision on aggregation location.

3.4.6.2.10 Element Reuse Strategy

The reusability of any of the transportation elements is a key driver in the design of the system. If additional follow-on missions to Mars are desired to establish routine access to Mars' surface, then the ability to reuse elements will be a key decision in enabling these missions. If reusability drives in-space transportation system mass, impacts may flow-down to the Earth launch campaign. Reuse strategy will include deciding whether to optimize the transportation system for all mission opportunities (enabling missions about every 2 years) or to optimize for other constraints (potentially limiting mission availability). Note that a 15-year service life is currently assumed for the Transit Habitat, enabling it to support dual roles as a Mars crew transport, as well as analog and lunar support missions.

3.4.6.3 System Concepts

Table 3-18. SAC22 In-Space Transportation Analysis Trade Space

Transportation Options / Mission Options	Hybrid Nuclear Electric / Chemical	Nuclear Thermal	Hybrid Solar Electric / Chemical	All-Chemical
Reference Mission 0 — Short Duration, Light Footprint, Short Surface Stay	**M0-NEP** 1.8 - 3.6 MW NEP; 3x 25k lb$_f$ LCH4/LO2; NRHO Assembly; NRHO Refueling; *SAC21*	**M0-NTP** 5x 12.5k lb$_f$ NTP; 10x 6mØ Drop Tanks; LEO Assembly; No Refueling	*M0-SEP* TBD ConOp; TBD Assembly; TBD Refueling	*M0-CP* TBD ConOp; TBD Assembly; TBD Refueling
Reference Mission 1 — Moderate Duration, Light Footprint, Short Surface Stay	**M1-NEP** <1.8 MW NEP; 3x 25k lb$_f$ LCH4/LO2; NRHO Assembly; NRHO Refueling; *SAC22*	**M1-NTP** 2x 12.5k lb$_f$ NTP; 5x 6mØ Drop Tanks; MEO Assembly; No Refueling; *SAC22*	**M1-SEP** 1MW Array / 400kW SEP; 3x 25k lb$_f$ LCH4/LO2; NRHO Assembly; NRHO Refueling; *SAC22*	**M1-CP** Tanker + Depot; Integrated Surface Payload; LEO Assembly; LEO Refueling; *SAC22*
Reference Mission 2 — Longer Duration, Minimum Energy, Light Footprint, Short Surface Stay	*M2-NEP* TBD ConOp; TBD Assembly; TBD Refueling	*M2-NTP* TBD ConOp; TBD Assembly; TBD Refueling	**M2-SEP** 700kW Array / 400kW SEP; 6x 1k lb$_f$ LCH4/LO2; NRHO Assembly; NRHO Refueling; *SAC22*	**M2-CP** Tanker + Depot; Integrated Surface Payload; LEO Assembly; LEO Refueling; *SAC22*

To better understand the performance of various propulsion system designs in the context of the analysis reference missions, four different propulsion and power options are currently under evaluation: a hybrid Nuclear Electric Propulsion (NEP)/Chemical Propulsion system, Nuclear Thermal Propulsion (NTP) System, hybrid Solar Electric Propulsion (SEP)/Chemical Propulsion System, and All-Chemical Propulsion Systems. The NEP/Chem hybrid and NTP systems are nuclear options, and the SEP/Chem and All-Chem systems are non-nuclear. Table 3-18 summarizes the various potential implementations of each system being analyzed, with respect to each of the three reference missions. Campaign manifest designations represent implementations with enough conceptual design fidelity for preliminary campaign assessments; implementations without such designations remain forward work.

Table 3-19. Mars Transportation System Functions and Example Concepts

Functions	Example Concept(s)		Heritage and Status
Transport crew from Earth vicinity to Mars vicinity and return	Transit Habitat		Reference conceptual design of an independent transit habitat informed by previous Next Space Technologies for Exploration Partnerships (NextSTEP) Appendix A activities
Transport crew system from Earth vicinity to Mars vicinity and return	Piloted Crewed Hybrid NEP/Chem Propulsion System		Power generation linked to terrestrial power systems, EP system potentially extensible from Gateway PPE
	Piloted Crewed Nuclear Thermal Propulsion System		Heritage to NERVA program
	Piloted Crewed Hybrid SEP/Chem Propulsion System		Reference concepts derived from Gateway's PPE, with extensibility to Mars NEP/Chem hybrid

Functions	Example Concept(s)		Heritage and Status
	Piloted Crewed Chemical Propulsion System		Concepts development ongoing, technology is relatively mature, but challenges remain.
Transport cargo from Earth vicinity to Mars vicinity	Cargo Chemical Propulsion System		Potential utilization of chemical stage of the hybrid NEP/Chem or SEP/Chem system.
	Cargo Nuclear Thermal Propulsion System		Similar to piloted variant, but potentially with fewer elements and/or lower thrust needs
	Cargo SEP/Chem Propulsion System		Similar to piloted variant, but potentially lower power needs
Crew Earth launch; reposition to DST; and Earth entry, descent, landing	Orion Spacecraft		Flight design Orion available for all mission opportunities for use with all transportation architecture variants.
Earth launch for cargo (> 5 m diameter; > 15,000 kg mass to translunar injection condition)	Commercial Heavy Lift Systems		Commercial heavy-lift conceptual designs available

Functions	Example Concept(s)		Heritage and Status
Earth launch for cargo >7 m diameter TBD kg mass to various aggregation orbits	Super Heavy Lift Systems		Space Launch System flight design available. Commercial super heavy lift conceptual designs in development
Aggregation and storage of propellant in space	Propellant Tanker Systems		Propellant tanker systems may not be needed for all architecture implementations. Concepts development is ongoing.
Provide systems and facilities to process, launch, and recover launch vehicles	Ground Systems		Government and commercial infrastructure available

As shown in Figure 3-16, Figure 3-17, Figure 3-18, and Figure 3-19, several conceptual designs of each transportation architecture are being developed to allow stakeholders to better assess option performance across the range of mission duration options. These figures demonstrate how vehicle size and complexity vary as just one parameter, total mission duration, is varied.

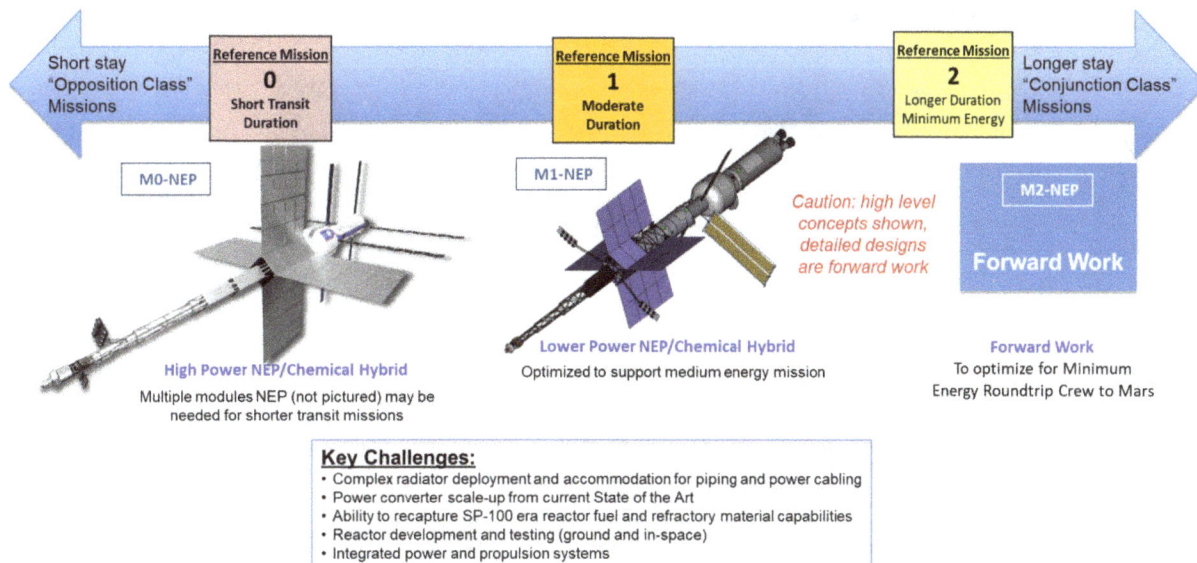

Key Challenges:
- Complex radiator deployment and accommodation for piping and power cabling
- Power converter scale-up from current State of the Art
- Ability to recapture SP-100 era reactor fuel and refractory material capabilities
- Reactor development and testing (ground and in-space)
- Integrated power and propulsion systems

Figure 3-16. Hybrid NEP/Chem Concepts Across a Range of Total Mission Durations

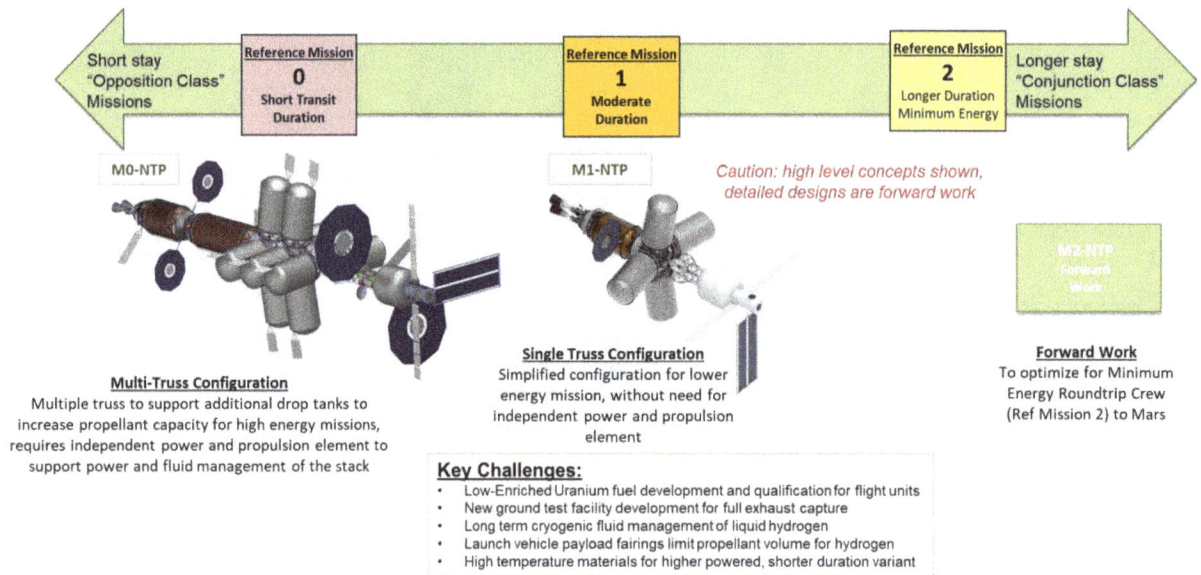

Figure 3-17. NTP Concepts Across a Range of Total Mission Durations

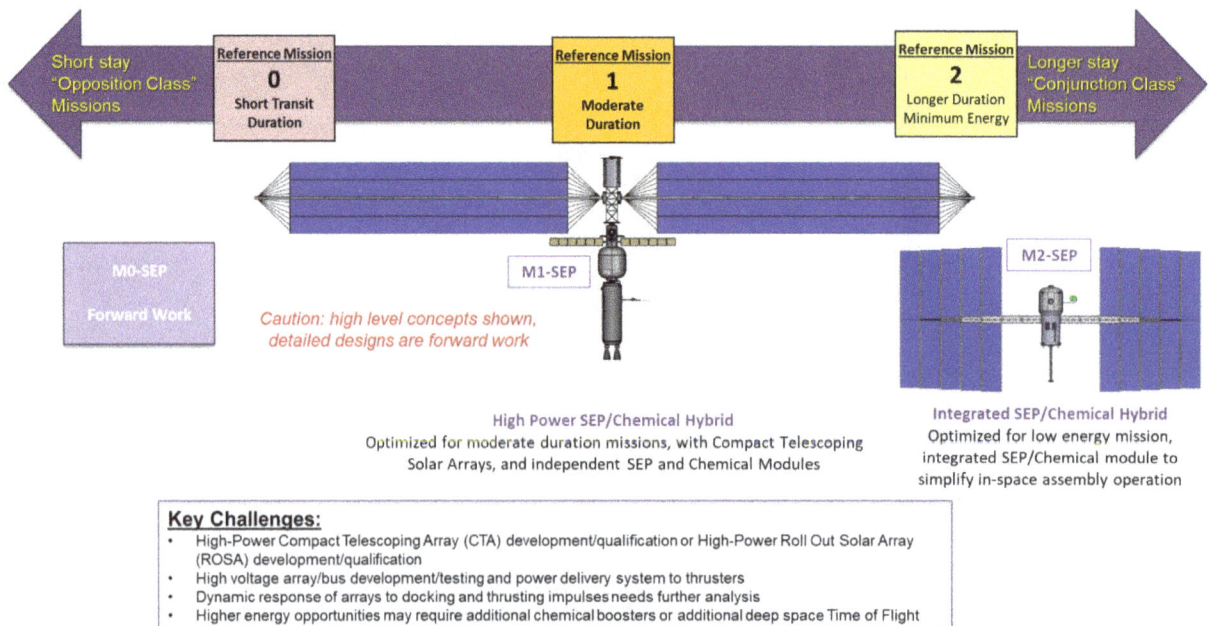

Figure 3-18. Hybrid SEP/Chem Concepts Across a Range of Total Mission Durations

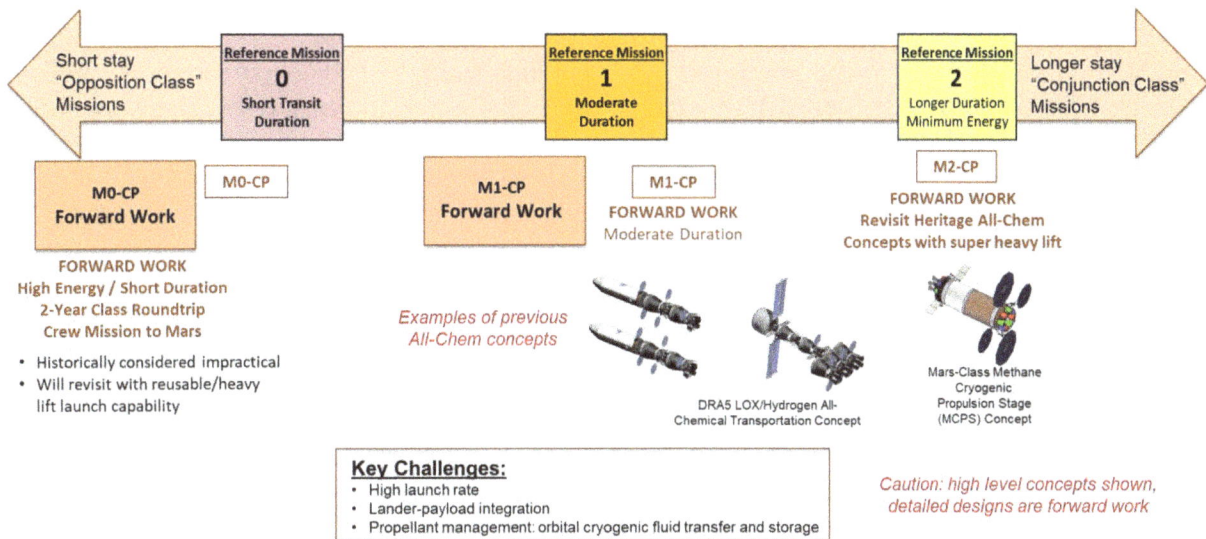

Figure 3-19. All-Chem Concepts Across a Range of Total Mission Durations

3.4.6.4　Concept of Operations

All four major transportation propulsion system architectures will require multiple Earth launches to an aggregation point, as well as in-space assembly, outfitting, and fueling of the deep space transportation system prior to crew boarding. However, the details of where, how, and when these steps occur vary by architecture, optimization choices made, and potential policy direction.

3.4.7　Mars Crew Support Systems

The Mars crew support architecture category covers elements needed to ensure that crew can perform across multiple mission phases. For the purpose of this document, it is assumed that these systems are common across all transportation architectures and surface concepts.

3.4.7.1　Communications

The Mars surface and close vicinity communications architecture is assumed to closely mirror the lunar architecture until further study is completed. A unique challenge for a Mars mission will be addressing the approximately two-week period during which the Sun interrupts the line-of-sight path between the crew and Earth, and no direct communication is possible. An uninterruptable relay could mitigate this blackout period, though it should be noted that this potentially increases the communications lag time, since the relay must be placed far enough from Mars to maintain line of sight to Earth when the Sun is between Earth and Mars. The relatively long- and variable-time delay for communications also poses a challenge for communications systems.

3.4.7.1.1　Functions

The Mars Communications System must transmit voice and data between Earth and the various Mars architecture vehicles (crew and cargo DST, MDS, MAV, and Orion), from Mars architecture vehicle to Mars architecture vehicle, from Mars surface to Mars orbit, and from surface asset to surface asset.

3.4.7.1.2 System Concepts

The Mars communications system elements will include communication components for EVA suits, mobility platforms, landers, transit vehicles, and uninterrupted Earth relay. Additional relay assets, if required, will be defined in future studies.

3.4.7.1.3 Concept of Operations

The Mars communications system concept of operations will be substantially different from lunar operations due to the delay caused by the increased distance from Earth-based ground support, up to 22 minutes each way, and the annual communications blackout of up to two weeks. The current architecture concepts posit the Mars transit vehicle acting as the primary relay between the surface systems network to Earth-based networks during the crewed surface phase of the mission. Another concept under consideration is having the surface crew relying on the orbital crew (that remain aboard the Mars transportation system) to provide low-latency verbal guidance and expertise to augment the longer latency-Earth support. It remains forward work to fully develop the Mars communications concept of operations that supports both near- and far-range operations with varying magnitudes of latency.

3.4.7.2 IVA and EVA Suits

3.4.7.2.1 Functions

The primary function of the Mars IVA suit system is to protect crew from the various environments encountered during Earth or Mars launch and landing. This includes potential contingencies such as cabin depressurization, fire, or toxic atmospheres. In addition, the IVA suit system must provide sufficient safety, mobility, communications, and comfort for crew to perform their duties inside a vehicle.

The primary function of the Mars EVA suit system is to protect crew from the various environments encountered during a Mars mission, independent of a pressurized cabin. This includes potential cabin depressurization or external vehicle contingency excursion events in the deep space transit environment, as well as nominal EVA operations on the Mars surface environment. In addition, the EVA suit system must provide sufficient safety, mobility, communications, and comfort for crew to perform their duties inside or outside a vehicle for time periods of up to a full workday. The Mars EVA suit must also integrate with Mars surface systems to enable safe, rapid crew ingress/egress from Mars surface system habitable volumes.

3.4.7.2.2 System Concepts

Table 3-20 summarizes the major IVA and EVA suit system concepts and status. The current Mars architecture assumes that IVA and EVA suit system elements used during Earth and Mars launch/entry and microgravity phases will be substantially like those used for similar operations at the Moon. However, the Mars surface suit, intended for Martian surface operations, will have some important differences from its lunar counterpart. The higher Martian gravity will make mass reduction a priority, and the thin Martian atmosphere will require changes in life support system operation. As noted above, habitable volumes available to at least the first Mars crew are likely to be different than those available to Artemis lunar crews, so ingress/egress strategy for the Mars crew cabins will influence Mars EVA suit design. Finally, planetary protection requirements for Mars are expected to be more stringent than on the Moon, which may influence permissible leakage rates or venting operations, as well as dust control techniques.

Table 3-20. Mars IVA/EVA System Functions and Example Concepts

Functions	Example Concept(s)		Heritage & Status
Crew life support and protection from the environment primarily intended for use inside a decompressed vehicle cabin	Launch and Landing IVA Suit		Orion Crew Survival System planned for use by Artemis.
Crew life support and protection from the environment for use inside a decompressed cabin, or for short excursions outside a vehicle in microgravity (not intended for planetary EVA)	Emergency EVA Suit		Reference concept available. May not be required if microgravity EVA suit and Mars surface suit can provide all necessary functionality
Crew life support and protection from the environment for longer excursions outside a vehicle during Earth-Mars transit	Microgravity EVA Suit		Reference conceptual design available. This concept may be substantially similar to EVA suits used on Gateway. May not be required if no nominal microgravity EVA is planned (default to emergency EVA suit in that case)
Crew life support and protection from the environment for excursions outside a vehicle on the Mars surface	Mars EVA Surface Suit		Reference conceptual design available. Can leverage Artemis lunar suit lessons learned, but the Mars suit will be unique due to gravity and environmental differences
Primarily safety equipment, such as tethers	Crew-worn EVA Accessories		Forward work

3.4.7.2.3 Concept of Operation

The Mars IVA suit system concept of operation is expected to be substantially similar to crew Earth launch/landing, lunar transit, and Gateway operations. Mars descent and ascent operations are expected to be similar enough to crew Earth launch/landing that a common IVA suit can be used for both. The Mars EVA suit system will be designed to allow crew members to perform autonomous and robotically assisted EVA exploration, research, construction, servicing, and repair operations in environments that exceed human capability. Current concepts assume the suit can egress and ingress habitable vehicles and provide life support, thermal control, protection from the environment, communications, and mobility/dexterity features designed to interact with spacecraft interfaces and supporting tools and equipment such that exploration, science, construction, and vehicle maintenance tasks can be done safely and effectively. Advanced concepts may also include designs that support rapid crew ingress / egress (to improve physical health outcomes associated with reduced pressure in suits and to reduce the risks of decompression sickness in crew), and enhance planetary protection protocols for the Mars environment or habitable vehicle and beyond (forward and backward contamination).

Current concepts assume that the pressure garment provides for resizing and modular component interchanges to enable proper fit across a wide range of anthropometries (1st-99th percentile). Interfaces in the Mars portable life support system and pressure garment are provided, which enable incremental upgrades to new technologies as the mission and destination evolve. Contamination from Mars dust constitutes a challenge to the design of mobility joints and the like, along with solutions to the introduction of surface contamination to crew habitat. The EVA crew will utilize advanced informatics designed into their suit system. These informatics will grant the EVA crew more autonomy with both tasks and suit monitoring due to the signal latency between Mars and Earth. It remains forward work to develop concepts and operations to address compatibility with the chemically reactive soil, as well as forward (planetary protection) and backward (crew health) contamination during crew ingress/egress operations. As one important link in breaking the chain of backward planetary protection, current operational concepts assume the Mars EVA surface suits are left behind on Mars, and crew members return to Mars orbit in their IVA suits.

3.4.7.3 Logistics Management

Requirements for Mars exploration missions often overlap with those driving the lunar architecture. Where possible Mars logistics management requirements should be met by leveraging these architectural similarities. In cases where options exist that are applicable to either Moon or Mars, M2M considerations should drive the selection of a concept that can satisfy both. However, some requirements will be specific to Mars and require a different approach.

3.4.7.3.1 Functions

The primary purpose of Mars logistics management is to provide for the transport, storage, tracking, and disposal of logistics, including crew consumables (food, clothing, etc.), life-support system commodities (breathing air, water, etc.), utilization, maintenance and spares, and other supplies and materials needed to implement crewed Mars missions. The logistics functions include coordination with in-space and surface transportation assets for performing support functions (either manually or with robotic assistants) needed to ensure that logistics arrive at the point of use as efficiently as possible. Functional capabilities assumed are detailed in the sections below.

3.4.7.3.2 Mars Transit Logistics Concept of Operations

Logistics for the in-space phase of the crewed mission are delivered and aggregated within the TH prior to Earth departure. Because of the amount of logistics required for the crewed Mars mission, the majority of the required logistics are delivered to the TH prior to the actual Mars crew departure. Logistics modules (standalone flights, SLS co-manifested or commercial co-manifested as needed) are used to both supply logistics for crew Gateway/TH operations in those years and to pre-emplace logistics for the crewed Mars mission.

A final logistics module is delivered to the TH in Earth orbit with the Mars crew immediately prior to the Mars mission. This final logistics module is used to deliver the logistics items that are most critical from a lifetime perspective, such as food, medicines, and crew-specific items. The logistic module is detached and disposed of, along with any trash, prior to TMI. Lower time priority items are delivered earlier to the TH.

The logistics stored on the TH prior to Earth departure include all of the logistics necessary to complete the mission. No pre-emplaced logistics are utilized to complete the in-space segment of the Mars mission. The total logistics include the logistics that are required for the crewed in-space duration of the mission, including time in Mars orbit for the entire crew. While nominally some or all of the crew will spend a portion of that time on the surface, TH logistics are still manifested for that period in case the crew are unable to complete the surface mission and must stay aboard the TH. Additional logistics are also manifested to cover the potential maximum crewed pre-departure duration. This period covers the potential Orion launch period, as well as an additional time to allow for logistics transfer and final mission preparations. This will allow Orion enough opportunities to get the crew to the spacecraft in the event of unexpected launch pad delays. Similarly, logistics are manifested to cover the end of mission Earth orbital duration, allowing for rendezvous and docking with Orion, as well as any time required for final transfers. If the entire Orion launch duration is not needed in the TH, any remaining consumables for that period are disposed of in the logistic module prior to TMI.

In addition to the nominal durations listed above, additional logistics are also manifested to cover contingency situations. Contingency gas and water are manifested to cover periods where regenerative ECLSS may be unavailable during repair activities. It is assumed that waste products are stored during this period and then processed after the repair is completed to build the contingency store back up. Additional contingency logistics may be manifested in the safe haven to cover period where the crew may be forced to shelter there.

Disposal of trash is a key issue for the in-space portion of the Mars mission. Because of the propulsive requirements, it is undesirable to accumulate trash in the TH. Methods to dispose of trash during the transits to and from Mars will be considered to reduce the total TH mass.

3.4.7.3.3 Mars Surface Logistics Concept of Operations

For surface operations, the logistics are pre-positioned in Mars orbit and then delivered to the Martian surface with the crew. Logistics delivery for surface operations is designed to reduce the burden on the crew and to preserve crew time for utilization activities. While logistics may be delivered either internally to the habitable surface elements or in external carriers, it is desirable to deliver the maximum possible amount of logistics in elements, directly accessible to the crew, reducing the need to transfer logistics from carriers.

Surface logistics are provided to support the entire surface missions. These include all required consumables and utilization, as well as any maintenance items that are required to provide high availability for surface systems. Logistics are also provided to cover various surface contingency scenarios. This could include consumables to provide protection against system failures, spares for systems, and additional consumables to cover extend duration, if necessary.

Disposal of trash is also a key issue for surface operations. Because of planetary protection considerations, it will be necessary to dispose of trash on the surface in a manner that prevents contamination on the landing site.

3.4.7.4 Crew Systems

Crew systems for habitability include direct crew care systems such as food and nutrition consumables and preparation equipment, personal hygiene systems including body waste management, clothing, housekeeping equipment and consumables, physiological countermeasure systems (such as aerobic and resistance exercise equipment), crew privacy systems and accommodations conducive for sleep.

Crew systems for in-flight medical operations include medical diagnostic and treatment equipment as well as consumables typically found in medical kits, while also ensuring that appropriate volume and restraints are available to support and safely restrain an injured or incapacitated crew member. Vehicle systems also need to support accessing and updating crew medical health records, accommodating private medical conferences, and sharing medical data with ground-based flight surgeons. A more detailed crew health concept of operations is typically found in a Crew Health Operations Concept document for each program.

Within the architecture, design for crew shall accommodate the physical characteristics, capabilities, and limitations of crew to ensure health, safety, and performance as well as continued hardware and system functionality. Physical characteristics may include anthropometry as well as range of motion, strength, and visual and hearing acuity. Behavioral capabilities include cognition and perception. Limitations for continued health may include radiation, acceleration and dynamic loads, acoustics, and vibrations, as well as environmental hazards such as thermal, atmospheric, or water. Accommodations may drive design for size of physical volumes, configuration of systems within, placement of restraints and mobility aids, accessibility of translation paths through spaces and hatches, and lighting to support tasks as well as emergencies and circadian alignment for sleep. For usability, durability, and maintenance and training minimization, systems design shall consider how and how often the human performs system tasking, typically planned for via early human system integration and demonstrated by task analysis and human-in-the-loop testing. Systems shall preclude injury to crew by design such as smoothing of sharp edges, elimination of pinch-points, prevention of unexpected energy release, electrical hazards, or chemical release, etc. Finally, systems shall be designed to account for crew survival scenarios for defined contingency scenarios.

As much as possible, these systems will be derived from Artemis and ISS systems, though the longer Mars mission duration, combined with limited resupply options, minimal spare parts, shelf-life limitations, and longer communications lag times will necessarily require modifications particularly to medical capabilities and food systems.

3.4.8 Open Questions, Ongoing Assessments, and Future Work

As noted, objective decomposition and use case and function definitions for the Mars segment have not been fully completed, and much of the trade space, particularly for sustained human presence, is still being assessed. Ongoing studies are evaluating return propellant strategies, surface infrastructure needs, and EDLA options to build integrated end-to-end campaign models, which in turn will support trade studies between the four candidate transportation technologies. Additional analysis remains to evaluate infrastructure and science objective implementation options, assess end-to-end architecture impacts, and develop integrated concepts of operation. Continuing human health and performance research is being assessed in the context of Mars mission durations and operational challenges, and risk mitigation options are being identified and evaluated. These examples are not intended to be a comprehensive list of open work; this document will be updated as additional analysis and research are identified and completed.

4.0   ASSESSMENT TO THE RECURRING TENETS

Within the Moon-to-Mars Strategy and Objectives Document, NASA established a high-level set of Recurring Tenets to guide the exploration architecture. These tenets give context to why and how NASA should conduct exploration missions. Using these objectives as a guide, the Moon-to-Mars Architecture and the elements will be managed and coordinated through a framework of sub-architectures and campaign segments to organize the decomposition. The essential nature of this framework is to ensure the progression of development towards greater objective satisfaction through campaign segments to return and sustain human presence in deep space. This constant traceability and iteration through the architecture process between the current state of execution and future goals and desired outcome will ensure infusion of technology, innovation, and emerging partners.

To ensure this progression and iteration of approach, assessments of progress and adherence to the Recurring Tenets is incorporated as an on-going process. The Moon-to-Mars architecture will be assessed against each Tenet, evaluating how these guiding principles are reflected in the current architecture. In addition, potential gaps in the current Moon-to-Mars architecture will also be identified to help guide future iteration and refinement of the architecture. These assessments will be coordinated with the stakeholders of each Tenet. The first of these initial assessments has been completed for RT-1 International Collaboration with inputs from the appropriate agency stakeholders, while the assessment for the other Tenets represents incomplete assessment awaiting future updates. Future revisions of this document will continue to update, evaluate, and assess the progress of the architecture in adhering to the Tenets.

4.1   RT-1 INTERNATIONAL COLLABORATION

RT-1: Partner with international community to achieve common goals and objectives.

Architecture Assessments:

An integral part of the Moon-to-Mars architecture is the desire to usher in a new era of exploration with the recognition of the mutual interest between NASA and international partners in the exploration and use of outer space for peaceful purposes. Coordination and cooperation between and among established and emerging actors in space is a foundational principle of Artemis. This is best accomplished through partnership and collaboration with members of the global community and should be reflected in every segment of the Moon-to-Mars Architecture.

NASA has many such international partnerships already in place, as well as ongoing discussions with prospective partners on potential contributions to the Moon-to-Mars architecture, such as the following:

- **European Service Module (ESM):** The European Space Agency (ESA) provided the ESM which powered the Orion spacecraft around the Moon on Artemis I and will do so on future Artemis missions.

- **Gateway:** ESA, the Japan Aerospace Exploration Agency (JAXA) and the Canadian Space Agency (CSA) are NASA's partners on the Gateway, all providing key elements to develop and operate this cislunar outpost.

- **Artemis I:** Several international partners provided payloads to research key knowledge gaps for deep space exploration. These partners included ESA, the German Aerospace Center (DLR), and the Israel Space Agency (ISA), which provided radiation experiments, and JAXA and the Italian Space Agency (ASI) which provided CubeSats.

- **Lunar science:** NASA's Science Mission Directorate is leading the CLPS initiative, to deliver payloads to cislunar orbit and the lunar surface, as well as partnering on international partner-led missions to achieve science, exploration, and technology development goals and priorities for the Moon. International partners are participating by joining US-led proposals to Payload and Research Investigation from the Surface of the Moon (PRISM) solicitations, being sponsored by NASA for delivery via CLPS, or contracting with a CLPS vendor directly for opportunities to reach the Moon. NASA has sponsored CLPS deliveries for payloads from ESA, the Korean Astronomy and Space Science Institute (KASI), CSA, and University of Bern, Switzerland. The French Centre National D'Études Spatiales (CNES) is participating in the US-led farside seismic payload and an electromagnetics experiment. NASA is also contributing instruments to various international missions, including: a laser retroreflector on JAXA's SLIM mission, an infrared imager on CSA's LEAP rover, a laser retroreflector on the Indian Space Research Organisation's (ISRO) Chandrayaan-3 mission, a radiation experiment on ISA's Beresheet-2 mission, and a neutron spectrometer on JAXA and ISRO's LuPEX mission. In addition, NASA contributed the ShadowCam, a camera that scans shadowed areas for ice deposits and landing zones, to the Korea Pathfinder Lunar Orbiter (KPLO), which was launched in 2022.

- **Space communications and navigation:** NASA is currently engaged in discussions with several space agencies regarding potential Artemis cooperation for ground stations, lunar relays, navigation assets, and lunar surface communications elements. These international partners include ESA, ASI, and JAXA, as well as the United Kingdom Space Agency, the New Zealand Space Agency, and the Australian Space Agency (ASA), among others. In addition, NASA is coordinating its planned commercial lunar relay procurement with a similar activity being conducted by ESA called Moonlight. Additionally, there is an effort to build a new ground station network of lunar antennas, including NASA-owned assets, commercial service providers, and international partner contributions; South Africa and Australia potentially may host one of these NASA-owned ground stations.

- **Additional Capability Discussions:** NASA is conducting technical studies and discussions with partners such as ESA, JAXA, ASI, CSA, the Luxembourg Space Agency, and the UAE, developing concepts in multiple areas including rovers, cargo delivery to the lunar surface, habitation, and lunar communications.

- **Technology:** International space technology partnerships generally focus on low technology readiness level and fundamental research, the results of which are then shared publicly. An example of this type of cooperation would involve dissimilar but redundant capabilities to augment common technology development objectives, such as NASA's ongoing collaboration with the Australian Space Agency regarding a mobile regolith collection capability on the lunar surface to support mutual ISRU demonstration objectives.

Architecture Gaps:

NASA envisions that a wide range of international partnerships will be enabled by the M2M architecture which will identify potential gaps and opportunities. Cooperation will occur across the full spectrum of opportunities–from infrastructure to science, technology, and education activities on and around the Moon, Mars, and beyond. New opportunities for cooperation will emerge as the architecture is further developed each year and collaboration is discussed between NASA and its prospective international partners. International cooperation will advance broad science, exploration, and space technology goals and objectives, as well any number of objectives related to education, inspiration, and public engagement.

It is important to recognize the iterative and ongoing nature of international discussions on architecture and our Moon-to-Mars partnerships. NASA has engaged and will continue engaging with international counterparts, both bilaterally and multilaterally. Along with bilateral discussions, some highlighted above, multilateral discussions have demonstrated utility in advancing mutual understanding and common exploration interests. Multilateral forums will continue to be utilized to further articulate NASA's exploration and science objectives with an eye toward further identifying areas of potential cooperation, such as the following multilateral efforts:

- The International Space Exploration Coordination Group (ISECG) is a coordination forum for interested space agencies to share their objectives and plans for exploration.

- The Lunar Exploration Analysis Group (LEAG) was established in 2004 to support NASA in providing analysis of scientific, technical, commercial, and operational issues in support of lunar exploration objectives and of their implications for lunar architecture planning and activity prioritization.

- The Mars Exploration Program Analysis Group (MEPAG) serves as a community-based, interdisciplinary forum for inquiry and analysis to support NASA Mars exploration objectives. MEPAG is responsible for providing the science input needed to plan and prioritize Mars exploration activities.

- The Solar System Exploration Research Virtual Institute (SSERVI) was formed to address questions fundamental to human and robotic exploration of the Moon, near Earth asteroids, the Martian moons Phobos and Deimos, and the near space environments of these target bodies. As a virtual institute, SSERVI funds investigators from a broad range of domestic institutions, bringing them together along with international partners via virtual technology to enable new scientific efforts.

- The International Mars Exploration Working Group (IMEWG) is a coalition of space agencies and institutions around the world that seeks to advance our collective human and robotic future on Mars. The International Space Life Sciences Working Group (ISLSWG) is a multilateral forum aimed at providing a more complete coordination of international development and use of spaceflight and special ground research facilities to enhance M2M objectives pertaining to space life sciences.

- The Interagency Operations Advisory Group (IOAG) provides a forum for identifying common needs across multiple agencies related to mission operations, space communications, and navigation interoperability.

In parallel, NASA and the Department of State will continue to engage with the community of Artemis Accords signatories to discuss the implementation of the principles of the Accords to ensure safe and sustainable space exploration.

4.2 RT-2 INDUSTRY COLLABORATION

Future assessment of architecture will be completed through coordination with appropriate stakeholders to provide a detailed analysis with respect to this tenet. A short summary of the current status is shown below.

RT-2: Partner with U.S. industry to achieve common goals and objectives.

Architecture Assessments:

NASA has long called upon the U.S. industrial base to provide the development and production of key exploration assets and provide foundational research to advance and enhance exploration capabilities. U.S. industry contributions will be critical throughout the Moon-to-Mars campaign segments. The elements included in the HLR segment are already leveraging commercial contributions. Future segments will continue this partnership. The Moon-to-Mars exploration architecture will depend on partnership with U.S. industry to provide exploration services in a sustainable and affordable manner.

Architecture Gaps:

U.S. industry contribution in future segments to enable exploration activities have not been fully captured or leveraged. With significant commercial interest in developing low-Earth orbit destinations in the near future, NASA needs to investigate opportunities to leverage and potentially supplement those investments to advance the state of knowledge of human space flight in support of the Moon-to-Mars architecture.

4.3   RT-3 CREW RETURN

Future assessment of architecture will be completed through coordination with appropriate stakeholders to provide a detailed analysis with respect to this tenet. A short summary of the current status is shown below.

RT-3:   Crew Return: Return crew safely to Earth while mitigating adverse impact to crew health.

Architecture Assessments:

In recognition of the inherent risks associated with human spaceflight, NASA treats the wellbeing and safe return of crews to be of paramount importance. Within the Moon-to-Mars architecture, this tenet is manifested across all campaign segments.

Architecture Gaps:

The Moon-to-Mars architecture treats the safety of the crew as an utmost concern. However, significant knowledge gaps exist relative to the adverse effects of long-term exposure to the deep space environment. The architecture will be developed to account for known health and medical concerns with deep space missions, as well as for contingency scenarios for failures in mission elements and systems. Significant knowledge has been gained in the on-going human health research aboard the International Space Station and will continue with lunar orbital and surface missions. Long duration Mars precursor missions conducted in cislunar space and on the lunar surface will provide some knowledge and experience. Likewise, knowledge of the reliability gaps with mission hardware, software, and operations will be tested and refined based on knowledge of LEO missions. However, these missions may not be sufficient to provide the necessary data to fully understand the risk associated with roundtrip missions to Mars. Furthermore, while the assets around cislunar space provide crew with safe haven and Earth-return capabilities, such capabilities may not exist for the Mars exploration crew, regardless of the architecture, and that lack remains a significant challenge to crew safety.

4.4   RT-4 CREW TIME

Future assessment of architecture will be completed through coordination with appropriate stakeholders to provide a detailed analysis with respect to this tenet. A short summary of the current status is shown below.

RT-4:   Maximize crew time available for science and engineering activities within planned mission duration.

Architecture Assessments:

One of the three pillars of the exploration strategy that is guiding the Moon-to-Mars architecture is the pursuit of scientific knowledge. Maximizing crew time is a critical driver for the exploration architecture across all segments of the campaign. Specifically, this refers to crew time made available for utilization activities, separate from other crew time allocations such as maintenance time. During the HLR campaign segment, the architecture and reference missions emphasize crew exploration by EVA on the lunar surface. This is enabled by allocating functions to the elements in this phase such that maintenance and construction overhead activities are minimized. Concurrently, utilization activities at the Gateway are conducted in cislunar space to complement the surface exploitation activities. The experiences gained in the exploration activities with the optimization of surface exploration missions on the Moon will guide the planning for the initial Mars surface mission to maximize the efficiency of crew time available for science and engineering activities.

Architecture Gaps:

There are knowledge gaps associated with the increasing exploration infrastructure and capabilities needed for the FE, SLE, and Humans to Mars segments. Although these increases are intended to result in a net benefit to available crew time, there is some uncertainty associated with the operational complexity, maintenance, and refurbishment demands they bring. Additional assessment is needed to bridge the knowledge gap to inform system design and operational planning.

4.5   RT-5 MAINTAINABILITY AND REUSE

Future assessment of architecture will be completed through coordination with appropriate stakeholders to provide a detailed analysis with respect to this tenet. A short summary of the current status is shown below.

RT-5:   When practical, design systems for maintainability, reuse, and/or recycling to support the long-term sustainability of operations and increase Earth independence.

Architecture Assessments:

To enable a safe, effective, and affordable architecture that achieves NASA's long-term exploration goals, the Moon-to-Mars architecture must be assessed to understand the implication of system maintainability, reuse, and/or recycling in support of long-term operation and increase Earth independence. Almost every element in the architecture is being designed to take advantage of some level of reuse, but understanding of risks associated with maintainability and reuse and their impact on safety, science, and long-term sustainability goals will be vital as the Moon-to-Mars architecture is refined and matured.

Architecture Gaps:

The maintainability of the exploration assets remains a major concern especially with the desire to maximize crew time for science and engineering activities. To extend the lifetime of any asset, a certain level of crew time must be dedicated for the maintenance of the systems. Additional assessment is needed to bridge the knowledge gap to inform system designs and operational planning. The reuse of elements will also likely require the refurbishment of elements and these activities are not well defined within the current architecture. Finally, system lifetime limitations from both a reuse and maintenance perspective must be evaluated.

4.6 RT-6 RESPONSIBLE USE

Future assessment of architecture will be completed through coordination with appropriate stakeholders to provide a detailed analysis with respect to this tenet. A short summary of the current status is shown below.

RT-6: Conduct all activities for the exploration and use of outer space for peaceful purposes consistent with international obligations, and principles for responsible behavior in space.

Architecture Assessments:

Architecture elements and reference missions across all campaign segments are being developed to adhere to existing law, policy, and guidance, including, but not limited to, established planetary protection policies. Additionally, the responsible use of the Moon-to-Mars architecture may require deeper scrutiny of cultural and societal implications of future exploration.

Architecture Gaps:

While NASA and its partners have made the commitment to ensure peaceful exploration and use of space, significant policy gaps exist with regards to the protection of future scientific and exploration needs. Planetary protection policies need to be refined and updated to ensure that they address the expansion of exploration to ensure the integrity of future exploration needs.

4.7 RT-7 INTEROPERABILITY

Future assessment of architecture will be completed through coordination with appropriate stakeholders to provide a detailed analysis with respect to this tenet. A short summary of the current status is shown below.

RT-7: Enable interoperability and commonality (technical, operations, and process standards) among systems, elements, and crews throughout the campaign.

Architecture Assessments:

The Moon-to-Mars architecture incorporates a diverse array of NASA programs with contributions from industry and international partners. Safely and successfully orchestrating the resulting array of systems requires a commitment to interoperability. Pre-formulation activities are working to establish a baseline level of interoperability across the exploration ecosystem, including in the areas of habitation, power, mobility, logistics, robotics, communications, and utilization. In most cases, baseline interfaces have been identified, but specific implementations are still being developed. To the greatest extent possible, these standardized interfaces are being developed to enable application to the later Humans to Mars Segment. The International Deep Space Interoperability Standards[16] will serve as the starting point for future interoperability assessments.

Architecture Gaps:

Interface standardization for all elements is being guided by the recurring tenets, but specific designs and requirements for interoperability are still being studied and refined. In many cases, these studies focus on a subset of relevant elements and may fall short of enterprise-wide coordination. Formal, coordinated functional analysis and standardization policies need to be developed to govern cross-program and cross-partner element development.

---

[16] International Deep Space Interoperability Standard. https://www.internationaldeepspacestandards.com

4.8 RT-8 LEVERAGE LOW-EARTH ORBIT

Future assessment of architecture will be completed through coordination with appropriate stakeholders to provide a detailed analysis with respect to this tenet. A short summary of the current status is shown below.

RT-8: Leverage infrastructure in low-Earth orbit to support Moon-to-Mars activities.

Architecture Assessments:

The success of the Moon-to-Mars architecture is dependent on leveraging the past and current human and robotic spaceflight experience to inform future system design and operational needs. The Moon-to-Mars architecture builds upon and leverages experience gained from past and present spaceflight experience. Specifically, the architecture relies on fundamental research that is currently on-going in LEO to support human health research that will be critical for both lunar and Mars missions. Component systems for Gateway have already begun testing in LEO and this integrated support will continue for both Gateway and the Mars transit habitat in the future. Future research, development, and testing are expected to leverage a combination of the International Space Station and other assets such as commercial free fliers and other partner-provided destinations.

Architecture Gaps:

Potential opportunities to leverage LEO infrastructure have not been fully explored, and additional studies and refinement are needed to evaluate all available options.

4.9 RT-9 COMMERCE AND SPACE DEVELOPMENT

Future assessment of architecture will be completed through coordination with appropriate stakeholders to provide a detailed analysis with respect to this tenet. A short summary of the current status is shown below.

RT-9: Foster the expansion of the economic sphere beyond Earth orbit to support U.S. industry and innovation.

Architecture Assessments:

Historically, the growth of the U.S. space industrial base has been both an enabler of exploration activities and a significant benefit to the broader economic sphere. This tenet responds to a number of U.S. space policy documents directing NASA to foster commercial industry and economic opportunity beyond Earth orbit. Partnerships are being considered for all elements and infrastructure to support the Moon-to-Mars Architecture, facilitating growth of a cislunar economic sphere in the future campaign segments. It is expected that the partnerships forged during the initial campaign segments will grow to incorporate a broader community and may have the potential to enable the formation of sustained human civilization beyond Earth. For example, the CLPS initiative will continue to utilize a service-based, competitive acquisition approach that enables rapid, affordable, and frequent access to the lunar surface.

Architecture Gaps:

Current plans for commercialization beyond LEO are in their formulation stage. Discussions across the industry and international partners will need to begin in earnest during the initial campaign segments to determine the desired future state and the evolutionary steps necessary to facilitate this vision. The intent of SLE is to address gaps with respect to RT-9 as future work.

APPENDIX A: FULL DECOMPOSITION OF LUNAR OBJECTIVES

This appendix shows the comprehensive decomposition for the lunar objectives into Characteristic and Needs, Use Cases, and Functions.

The Human Lunar Return Segments that have Use Cases and Functions mapped to specific elements (as described in Section 3.1) are highlighted as purple text.

A.1 SCIENCE

A.1.1 Lunar/Planetary Science

*Mapped to HLR Segment

ID	Functions	ID	Use Cases	← Characteristics & Needs	← Objectives	ID
F-13	Transport crew and systems from cislunar space to lunar surface South Pole sites	UC-21	Crewed missions to distributed landing sites around South Pole	←Visit geographically diverse sites around the South Pole and non-polar regions	Uncover the record of solar system origin and early history, by determining how and when planetary bodies formed and differentiated, characterizing the impact chronology of the inner solar system as recorded on the Moon and Mars, and characterize how impact rates in the inner solar system have changed over time as recorded on the Moon and Mars.	LPS-1
F-44	Transport crew and systems from cislunar space to lunar surface non-polar sites	UC-22	Crewed missions to non-polar landing sites, including the lunar far side	←Identify and collect samples from multiple locations, including frozen samples from PSRs, on the lunar surface		
F-40	Provide robotic systems on lunar surface controlled from Earth and/or cislunar space	UC-23	Robotic survey of potential crewed landing sites to identify locations of interest	←Return a variety of types of samples collected from the lunar surface, including of soil, pebbles, and rocks (hand sample size)		
F-45	Provide robotic systems in PSRs on lunar surface controlled from Earth and/or cislunar space	UC-48	Robotic survey of PSRs near potential crewed landing sites to identify locations of interest	←Return a variety of samples from the lunar subsurface		
F-24	Provide PNT capability on the lunar surface	UC-24	Crew excursions to locations distributed around landing site	←Return select samples of regolith, rock, and subsurface materials in containers sealed in lunar vacuum		
F-30	Provide local unpressurized crew surface mobility			←Emplace and operate science packages at distributed sites on the lunar surface		
F-31	Provide pressurized crew surface mobility			–Ability to know where samples are collected from		
F-28	Conduct crew surface EVA activities	UC-25	Crew EVA exploration and identification of samples	←Deploy scientific payloads at distances outside the blast zone of the ascent vehicle		
F-23	Provide high bandwidth, high availability comms between lunar surface and Earth			←Provide power to deployed science payloads to enable sustained operation for durations of several years		
F-46	Crew survey of areas of interest and identification of samples					
F-24	Provide PNT capability on the lunar surface					
F-28	Conduct crew surface EVA activities	UC-26				

*Mapped to HLR Segment

ID	Functions	ID	Use Cases	← Characteristics & Needs	← Objectives	ID
F-32	Recover and package surface samples		Crew collection of samples from lighted areas	←Ability for the Science Evaluation Room to exchange data and interact with crew in real time		
F-46	Crew survey of areas of interest and identification of samples					
F-47	Provide tools and containers to recover and package surface samples					
F-48	Store collected samples on lunar surface					
F-42	Transport cargo from lunar surface to Earth	UC-20	Return of collected samples to Earth in sealed sample containers			
F-43	Recover samples after splashdown					
F-49	Transport cargo from lunar surface to Earth at ambient/native temperature	UC-97	Return of collected samples to Earth at ambient/native temperatures in sealed sample containers			
F-121	Recover conditioned samples at ambient/native temperature after splashdown					
F-50	Orbital observation and sensing of lunar surface	UC-27	Orbital surveys before, during, and after crew missions			
F-28	Conduct crew surface EVA activities	UC-28	Crew emplacement and set-up of science packages on lunar surface w/ long-term remote operation			
F-18	Transport cargo from Earth to lunar surface					
F-51	Provide surface power for deployed payloads					
F-13	Transport crew and systems from cislunar space to lunar surface South Pole sites	UC-21	Crewed missions to distributed landing sites around South Pole	←Visit geographically diverse sites around the South Pole and non-polar regions	Advance understanding of the geologic processes that affect planetary bodies by determining the interior structures, characterizing the magmatic histories, characterizing ancient, modern, and evolution of atmospheres/exospheres, and investigating how active processes modify the surfaces of the Moon and Mars.	LPS-2
F-44	Transport crew and systems from cislunar space to lunar surface non-polar sites	UC-29	Crewed missions to non-polar landing sites	←Identify and collect samples from multiple locations, including frozen samples from PSRs, on the lunar surface		
F-40	Provide robotic systems on lunar surface controlled from Earth and/or cislunar space	UC-23	Robotic survey of potential crewed landing sites to identify locations of interest	←Return a variety of types of samples collected from the lunar surface, including of soil, pebbles, and rocks (hand sample size)		
F-45	Provide robotic systems in PSRs on lunar surface controlled from Earth and/or cislunar space	UC-48	Robotic survey of PSRs near potential crewed landing sites to identify locations of interest	←Return a variety of samples from the lunar subsurface		
F-30	Provide local unpressurized crew surface mobility	UC-24	Crew excursions to locations distributed around landing site	←Return select samples of regolith, rock, and subsurface materials in containers sealed in lunar vacuum		
F-31	Provide pressurized crew surface mobility			←Emplace and operate science packages at distributed sites on the lunar surface		
F-24	Provide PNT capability on the lunar surface					

*Mapped to HLR Segment

ID	Functions	ID	Use Cases	← Characteristics & Needs	← Objectives	ID
F-23	Provide high bandwidth, high availability comms between lunar surface and Earth	UC-25	Crew EVA exploration and identification of samples	←Ability to know where samples are collected from ←Deploy scientific payloads at distances outside the blast zone of the ascent vehicle ←Provide power to deployed science payloads to enable sustained operation for durations of several years ←Ability for the Science Evaluation Room to exchange data and interact with crew in real time		
F-24	Provide PNT capability on the lunar surface					
F-46	Crew survey of areas of interest and identification of samples					
F-28	Conduct crew surface EVA activities					
F-46	Crew survey of areas of interest and identification of samples	UC-26	Crew collection of samples from lighted areas			
F-28	Conduct crew surface EVA activities					
F-32	Recover and package surface samples					
F-47	Provide tools and containers to recover and package surface samples					
F-48	Store collected samples on lunar surface					
F-42	Transport cargo from lunar surface to Earth	UC-20	Return of collected samples to Earth in sealed sample containers			
F-43	Recover samples after splashdown					
F-49	Transport cargo from lunar surface to Earth at ambient/native temperature	UC-97	Return of collected samples to Earth at ambient/native temperatures in sealed sample containers			
F-121	Recover conditioned samples at ambient/native temperature after splashdown					
F-50	Orbital observation and sensing of lunar surface	UC-27	Orbital surveys before, during, and after crew missions			
F-28	Conduct crew surface EVA activities	UC-28	Crew emplacement and set-up of science packages on lunar surface w/ long-term remote operation			
F-18	Transport cargo from Earth to lunar surface					
F-51	Provide surface power for deployed payloads					
F-52	Conduct crew surface EVA activities into PSRs	UC-30	Crew/robotic collection of samples from PSRs			
F-53	Crew and/or robotic survey of areas of interest and identification of samples in PSRs					
F-54	Recover and package surface samples in PSRs					
F-55	Store collected samples on lunar surface at ambient/native temperature					

*Mapped to HLR Segment

ID	Functions	ID	Use Cases	← Characteristics & Needs	← Objectives	ID
F-13	Transport crew and systems from cislunar space to lunar surface South Pole sites	UC-21	Crewed missions to distributed landing sites around South Pole	←Visit geographically diverse sites around the South Pole and non-polar regions ←Identify and collect samples from multiple locations, including frozen samples from PSRs, on the lunar surface ←Return a variety of types of samples collected from the lunar surface, including of soil, pebbles, and rocks (hand sample size) ←Return a variety of samples from the lunar subsurface ←Return select samples of regolith, rock, and subsurface materials in containers sealed in lunar vacuum ←Emplace and operate science packages at distributed sites on the lunar surface ←Ability to know where samples are collected from ←Deploy scientific payloads at distances outside the blast zone of the ascent vehicle ←Provide power to deployed science payloads to enable sustained operation for durations of several years ←Ability for the Science Evaluation Room to exchange data and interact with crew in real time	Reveal inner solar system volatile origin and delivery processes by determining the age, origin, distribution, abundance, composition, transport, and sequestration of lunar and Martian volatiles.	LPS-3
F-44	Transport crew and systems from cislunar space to lunar surface non-polar sites	UC-29	Crewed missions to non-polar landing sites			
F-40	Provide robotic systems on lunar surface controlled from Earth and/or cislunar space	UC-23	Robotic survey of potential crewed landing sites to identify locations of interest			
F-45	Provide robotic systems in PSRs on lunar surface controlled from Earth and/or cislunar space	UC-48	Robotic survey of PSRs near potential crewed landing sites to identify locations of interest			
F-30	Provide local unpressurized crew surface mobility	UC-24	Crew excursions to locations distributed around landing site			
F-31	Provide pressurized crew surface mobility					
F-24	Provide PNT capability on the lunar surface					
F-23	Provide high bandwidth, high availability comms between lunar surface and Earth	UC-25	Crew EVA exploration and identification of samples			
F-28	Conduct crew surface EVA activities					
F-24	Provide PNT capability on the lunar surface					
F-28	Conduct crew surface EVA activities	UC-26	Crew collection of samples from lighted areas			
F-32	Recover and package surface samples					
F-46	Crew survey of areas of interest and identification of samples					
F-47	Provide tools and containers to recover and package surface samples					
F-48	Store collected samples on lunar surface					
F-42	Transport cargo from lunar surface to Earth	UC-20	Return of collected samples to Earth in sealed sample containers			
F-43	Recover samples after splashdown					
F-49	Transport cargo from lunar surface to Earth at ambient/native temperature	UC-97	Return of collected samples to Earth at ambient/native temperatures in sealed sample containers			
F-121	Recover conditioned samples at ambient/native temperature after splashdown					

*Mapped to HLR Segment

ID	Functions	ID	Use Cases	← Characteristics & Needs	← Objectives	ID
F-50	Orbital observation and sensing of lunar surface	UC-27	Orbital surveys before, during, and after crew missions			
F-28	Conduct crew surface EVA activities	UC-28	Crew emplacement and set-up of science packages on lunar surface w/ long-term remote operation			
F-18	Transport cargo from Earth to lunar surface					
F-51	Provide surface power for deployed payloads					
F-52	Conduct crew surface EVA activities into PSRs	UC-30	Crew/robotic collection of samples from PSRs			
F-53	Crew and/or robotic survey of areas of interest and identification of samples in PSRs					
F-54	Recover and package surface samples in PSRs					
F-55	Store collected samples on lunar surface at ambient/native temperature					

A.1.2 Heliophysics Science

*Mapped to HLR Segment

ID	Functions	ID	Use Cases	← Characteristics & Needs	← Objectives	ID
F-56	Deliver utilization cargo to cislunar elements	UC-31	Crew emplacement and set-up of Heliophysics packages at cislunar elements w/ long-term remote operation	←Emplace and operate science instrumentation in a variety of lunar orbits	Improve understanding of space weather phenomena to enable enhanced observation and prediction of the dynamic environment from space to the surface at the Moon and Mars.	HS-1
F-57	External mounting points on cislunar elements			←Emplace and operate science instrumentation for solar monitoring off the Earth-Sun line		
F-58	Deliver free-flyers to cislunar space	UC-32	Autonomous deployment and long-term operation of free-flying packages in various lunar orbits	←Emplace and operate science instrumentation on the lunar surface		
F-28	Conduct crew surface EVA activities	UC-33	Crew emplacement and set-up of Heliophysics packages on lunar surface w/ long-term remote operation	←Provide power, communications, and data to deployed science payloads to enable sustained operation for durations of several years		
F-18	Transport cargo from Earth to lunar surface					
F-40	Provide robotic systems on lunar surface controlled from Earth and/or cislunar space	UC-23	Robotic survey of potential crewed landing sites to identify locations of interest	←Document and collect samples from multiple locations on the lunar surface	Determine the history of the Sun and solar system as recorded in the lunar and Martian regolith.	HS-2
F-30	Provide local unpressurized crew surface mobility	UC-24	Crew excursions to locations distributed around landing site	←Return to Earth drill core samples of lunar soil collected from the lunar subsurface at a variety of depths		
F-31	Provide pressurized crew surface mobility					

*Mapped to HLR Segment

ID	Functions	ID	Use Cases	← Characteristics & Needs	← Objectives	ID
F-24	Provide PNT capability on the lunar surface			←Emplace and operate science instrumentation on the lunar surface		
F-28	Conduct crew surface EVA activities					
F-24	Provide PNT capability on the lunar surface					
F-46	Crew survey of areas of interest and identification of samples	UC-25	Crew EVA exploration and identification of samples			
F-23	Provide high bandwidth, high availability comms between lunar surface and Earth					
F-32	Recover and package surface samples	UC-34	Crew collection of regolith samples from a variety of sites			
F-48	Store collected samples on lunar surface					
F-42	Transport cargo from lunar surface to Earth	UC-20	Return of collected samples to Earth in sealed sample containers			
F-43	Recover samples after splashdown					
F-28	Conduct crew surface EVA activities	UC-33	Crew emplacement and set-up of Heliophysics packages on lunar surface w/ long-term remote operation			
F-18	Transport cargo from Earth to lunar surface					
F-13	Transport crew and systems from cislunar space to lunar surface South Pole sites	UC-21	Crewed missions to distributed landing sites around South Pole		Investigate and characterize fundamental plasma processes, including dust-plasma interactions, using the cislunar, near-Mars, and surface environments as laboratories.	HS-3
F-44	Transport crew and systems from cislunar space to lunar surface non-polar sites	UC-29	Crewed missions to non-polar landing sites	←Emplace and operate science instrumentation in a variety of lunar orbits		
F-33	Transport cargo from Earth to elements in deep space	UC-31	Crew emplacement and set-up of Heliophysics packages at cislunar elements w/ long-term remote operation	←Emplace and operate science instrumentation in globally distributed locations on the lunar surface		
F-58	Deliver free-flyers to cislunar space	UC-32	Autonomous deployment and long-term operation of free-flying packages in various lunar orbits	←Document and collect samples from globally distributed locations on the lunar surface		
F-28	Conduct crew surface EVA activities	UC-33	Crew emplacement and set-up of Heliophysics packages on lunar surface w/ long-term remote operation			
F-18	Transport cargo from Earth to lunar surface					
F-28	Conduct crew surface EVA activities	UC-26	Crew collection of samples from lighted areas			
F-46	Crew survey of areas of interest and identification of samples					
F-32	Recover and package surface samples					

*Mapped to HLR Segment

ID	Functions	ID	Use Cases	← Characteristics & Needs	← Objectives	ID
F-47	Provide tools and containers to recover and package surface samples					
F-48	Store collected samples on lunar surface					
F-56	Deliver utilization cargo to cislunar elements	UC-31	Crew emplacement and set-up of Heliophysics packages at cislunar elements w/ long-term remote operation			
F-57	External mounting points on cislunar elements					
F-58	Deliver free-flyers to cislunar space	UC-32	Autonomous deployment and long-term operation of free-flying packages in various lunar orbits	←Emplace and operate science packages in lunar orbit	Improve understanding of magnetotail and pristine solar wind dynamics in the vicinity of the Moon and around Mars.	HS-4
F-28	Conduct crew surface EVA activities	UC-33	Crew emplacement and set-up of Heliophysics packages on lunar surface w/ long-term remote operation	←Emplace and operate science instrumentation on the lunar surface		
F-18	Transport cargo from Earth to lunar surface					

A.1.3 Human and Biological Science

*Mapped to HLR Segment

ID	Functions	ID	Use Cases	← Characteristics & Needs	← Objectives	ID
F-34	Provide pressurized, habitable environment in deep space	UC-35	Analog missions with extended durations in NRHO, followed by lunar surface missions	←Provide numerous mid-duration crew increments on the lunar surface	Understand the effects of short- and long-duration exposure to the environments of the Moon, Mars, and deep space on biological systems and health, using humans, model organisms, systems of human physiology, and plants.	HBS-1
F-33	Transport cargo from Earth to elements in deep space					
F-35	Move cargo into habitable elements in deep space			←Provide numerous long duration crew increments in cis lunar space prior to surface mission		
F-39	Provide crew remote medical systems in cislunar space					
F-26	Provide crew remote medical systems on lunar surface			←Conduct crew transitions from micro-gravity to partial gravity		
F-59	Autonomous crew landing on lunar surface					
F-60	Provide IVA laboratory space on lunar surface	UC-36	IVA facilities (e.g., instruments, racks, stowage, power) on the lunar surface to enable biological science analyses			
F-48	Store collected samples on lunar surface					

*Mapped to HLR Segment

ID	Functions	ID	Use Cases	← Characteristics & Needs	← Objectives	ID
F-61	Transport cargo from lunar surface to Earth at freezing temperatures					
F-62	Provide IVA laboratory space in cislunar space	UC-37	IVA facilities (e.g., instruments, racks, stowage, power) in cislunar orbit to enable biological science analyses			
F-63	Aggregate and extended storage collected samples in cislunar space					
F-64	Transport cargo from cislunar space to Earth at freezing temperatures					
F-34	Provide pressurized, habitable environment in deep space	UC-38	Analog missions with extended durations in NRHO, followed by lunar surface missions, with autonomous/semi-autonomous operations	←Provide numerous mid-duration crew increments on the lunar surface ←Provide numerous long duration crew increments in cis lunar space prior to surface mission ←Conduct crew transitions from micro-gravity to partial gravity	Evaluate and validate progressively Earth-independent crew health & performance systems and operations with mission durations representative of Mars-class missions.	HBS-2
F-33	Transport cargo from Earth to elements in deep space					
F-35	Move cargo into habitable elements in deep space					
F-39	Provide crew remote medical systems in cislunar space					
F-26	Provide crew remote medical systems on lunar surface					
F-59	Autonomous crew landing on lunar surface					
F-34	Provide pressurized, habitable environment in deep space	UC-35	Analog missions with extended durations in NRHO, followed by lunar surface missions	←Provide numerous mid-duration crew increments on the lunar surface ←Provide numerous long-duration crew increments in cis lunar space prior to surface mission ←Conduct crew transitions from micro-gravity to partial gravity	Characterize and evaluate how the interaction of exploration systems and the deep space environment affect human health, performance, and space human factors to inform future exploration-class missions.	HBS-3
F-33	Transport cargo from Earth to elements in deep space					
F-35	Move cargo into habitable elements in deep space					
F-39	Provide crew remote medical systems in cislunar space					
F-26	Provide crew remote medical systems on lunar surface					
F-59	Autonomous crew landing on lunar surface					
F-60	Provide IVA laboratory space on lunar surface	UC-36	IVA facilities (e.g., instruments, racks, stowage, power) on the lunar surface to enable biological science analyses			
F-48	Store collected samples on lunar surface					
F-61	Transport cargo from lunar surface to Earth at freezing temperatures					

*Mapped to HLR Segment

ID	Functions	ID	Use Cases	← Characteristics & Needs	← Objectives	ID
F-62	Provide IVA laboratory space in cislunar space	UC-37	IVA facilities (e.g., instruments, racks, stowage, power) in cislunar orbit to enable biological science analyses			
F-63	Aggregate and extended storage collected samples in cislunar space					
F-64	Transport cargo from cislunar space to Earth at freezing temperatures					

A.1.4 Physics and Physical Science

*Mapped to HLR Segment

ID	Functions	ID	Use Cases	← Characteristics & Needs	← Objectives	ID
F-65	Transport crew and systems from cislunar space to lunar surface far side sites	UC-39	Crew emplacement and set-up of astrophysics instrumentation on far side lunar surface w/ long-term remote operation	←Emplace and operate astrophysics science instrumentation on the far side lunar surface ←IVA facilities (e.g., instruments, racks, stowage, power) on the lunar surface to enable fundamental physics experiments	Conduct astrophysics and fundamental physics investigations of deep space and deep time from the radio quiet environment of the lunar far side.	PPS-1
F-66	Transport cargo to the far side lunar surface					
F-67	Conduct crew far side surface EVA activities					
F-62	Provide IVA laboratory space in cislunar space	UC-40	Crew conduct fundamental physics experiments while in the habitable volume on the lunar surface			
F-33	Transport cargo from Earth to elements in deep space	UC-41	Crew emplacement and set-up of fundamental physics experiments at cislunar elements w/ long-term remote operation	←Emplace and operate science packages in lunar orbit ←Emplace and operate science packages on the lunar surface ←Provide IVA laboratory space on the lunar surface and provide for crew time to conduct experiments	Advance understanding of physical systems and fundamental physics by utilizing the unique environments of the Moon, Mars, and deep space.	PPS-2
F-18	Transport cargo from Earth to lunar surface	UC-42	Crew emplacement and set-up of fundamental physics experiments on lunar surface w/ long-term remote operation			
F-28	Conduct crew surface EVA activities					
F-60	Provide IVA laboratory space on lunar surface	UC-43	Crew IVA research in dedicated science laboratory on the lunar surface			
F-42	Transport cargo from lunar surface to Earth	UC-44	Crew stows samples collected during fundamental physics experiments for return to Earth			

A.1.5 Science Enabling

*Mapped to HLR Segment

ID	Functions	ID	Use Cases	← Characteristics & Needs	← Objectives	ID
F-68	Provide training of crew prior to mission	UC-45	Provide advanced geology training as well as detailed objective-specific training to astronauts for science activities prior to each Artemis mission	←Train astronauts to be field geologists and to perform additional science tasks during Artemis missions, through field and classroom training	Provide in-depth, mission-specific science training for astronauts to enable crew to perform high-priority or transformational science on the surface of the Moon, and Mars, and in deep space.	SE-1
F-69	Provide in situ training of crew in cislunar space	UC-46	Train astronauts for science tasks during an Artemis mission utilizing in situ training capabilities			
F-70	Provide in situ training of crew on lunar surface					
F-23	Provide high bandwidth, high availability comms between lunar surface and Earth	UC-47	Allow ground personnel and science team to directly engage with astronauts on the surface and in lunar orbit, augmenting the crew's effectiveness at conducting science activities.	←Implement communications systems to enable high bandwidth, high availability communications between Earth-based personnel, surface crew, and science packages on the surface ←Train Earth-based scientists, integrated with FOD, to support crew activities in real time	Enable Earth-based scientists to remotely support astronaut surface and deep space activities using advanced techniques and tools.	SE-2
F-71	Provide high bandwidth, high availability comms between cislunar space and Earth					
F-45	Provide robotic systems in PSRs on lunar surface controlled from Earth and/or cislunar space	UC-48	Robotic survey of PSRs near potential crewed landing sites to identify locations of interest	←Return to Earth frozen samples from a variety of depths in their pristine state to JSC curation facilities ←Return select samples of soil, rock, and subsurface materials in containers sealed in lunar vacuum to JSC curation facilities ←Employ tools, including temperature sensors, to support acquisition of frozen samples, manufactured in accordance with science requirements to minimize sample contamination ←Employ sample containers appropriate for the specimens collected and science needs (e.g. contamination considerations), including sealed containers and drill core tubes	Develop the capability to retrieve core samples of frozen volatiles from permanently shadowed regions on the Moon and volatile-bearing sites on Mars and to deliver them in pristine states to modern curation facilities on Earth.	SE-3
F-72	Provide local unpressurized crew surface mobility into PSRs	UC-49	Crew excursions to PSRs near landing site			
F-31	Provide pressurized crew surface mobility					
F-28	Conduct crew surface EVA activities	UC-25	Crew EVA exploration and identification of samples			
F-46	Crew survey of areas of interest and identification of samples					
F-23	Provide high bandwidth, high availability comms between lunar surface and Earth					
F-24	Provide PNT capability on the lunar surface					
F-28	Conduct crew surface EVA activities	UC-26	Crew collection of samples from lighted areas			
F-46	Crew survey of areas of interest and identification of samples					
F-32	Recover and package surface samples					
F-47	Provide tools and containers to recover and package surface samples					

*Mapped to HLR Segment

ID	Functions	ID	Use Cases	← Characteristics & Needs	← Objectives	ID
F-48	Store collected samples on lunar surface					
F-52	Conduct crew surface EVA activities into PSRs	UC-30	Crew/robotic collection of samples from PSRs			
F-53	Crew and/or robotic survey of areas of interest and identification of samples in PSRs					
F-54	Recover and package surface samples in PSRs					
F-55	Store collected samples on lunar surface at ambient/native temperature					
F-49	Transport cargo from lunar surface to Earth at ambient/native temperature	UC-97	Return of collected samples to Earth at ambient/native temperatures in sealed sample containers			
F-121	Recover conditioned samples at ambient/native temperature after splashdown					
F-42	Transport cargo from lunar surface to Earth	UC-20	Return of collected samples to Earth in sealed sample containers			
F-43	Recover samples after splashdown					
F-40	Provide robotic systems on lunar surface controlled from Earth and/or cislunar space	UC-23	Robotic survey of potential crewed landing sites to identify locations of interest	←Visit geologically diverse sites around the South Pole and non-polar regions ←Document and collect samples from multiple locations, including frozen samples from PSRs, on the lunar surface ←Return to Earth a variety of types of samples collected from the lunar surface, including of soil, pebbles, and rocks (hand sample size) ←Return to Earth a variety of samples from the lunar subsurface at a variety of depths ←Employ tools to support acquisition of samples, including soil, pebbles, hand-sized rock samples, and drill cores, manufactured in accordance with science requirements to minimize sample contamination	Return representative samples from multiple locations across the surface of the Moon and Mars, with sample mass commensurate with mission-specific science priorities.	SE-4
F-45	Provide robotic systems in PSRs on lunar surface controlled from Earth and/or cislunar space	UC-48	Robotic survey of PSRs near potential crewed landing sites to identify locations of interest			
F-24	Provide PNT capability on the lunar surface	UC-24	Crew excursions to locations distributed around landing site			
F-30	Provide local unpressurized crew surface mobility					
F-31	Provide pressurized crew surface mobility					
F-28	Conduct crew surface EVA activities	UC-25	Crew EVA exploration and identification of samples			
F-46	Crew survey of areas of interest and identification of samples					
F-23	Provide high bandwidth, high availability comms between lunar surface and Earth					
F-24	Provide PNT capability on the lunar surface					
F-28	Conduct crew surface EVA activities	UC-26				

*Mapped to HLR Segment

ID	Functions	ID	Use Cases	← Characteristics & Needs	← Objectives	ID
F-46	Crew survey of areas of interest and identification of samples		Crew collection of samples from lighted areas	←Employ sample containers appropriate for the specimens collected and science needs (e.g. contamination considerations), including bags, sealed containers, and drill core tubes		
F-32	Recover and package surface samples					
F-47	Provide tools and containers to recover and package surface samples					
F-48	Store collected samples on lunar surface					
F-52	Conduct crew surface EVA activities into PSRs	UC-30	Crew/robotic collection of samples from PSRs			
F-53	Crew and/or robotic survey of areas of interest and identification of samples in PSRs					
F-54	Recover and package surface samples in PSRs					
F-55	Store collected samples on lunar surface at ambient/native temperature					
F-49	Transport cargo from lunar surface to Earth at ambient/native temperature	UC-97	Return of collected samples to Earth at ambient/native temperatures in sealed sample containers			
F-121	Recover conditioned samples at ambient/native temperature after splashdown					
F-42	Transport cargo from lunar surface to Earth	UC-20	Return of collected samples to Earth in sealed sample containers			
F-43	Recover samples after splashdown					
F-40	Provide robotic systems on lunar surface controlled from Earth and/or cislunar space	UC-50	Utilize robots to survey sites around the South Pole.	←Robotic surveys of potential landing sites, including video and in situ measurements	Use robotic techniques to survey sites, conduct in-situ measurements, and identify/stockpile samples in advance of and concurrent with astronaut arrival, to optimize astronaut time on the lunar and Martian surface and maximize science return.	SE-5
F-73	Provide robotic systems on lunar surface controlled from Earth and/or cislunar space b	UC-51	Utilize robots to take measurements and conduct experiments on the surface	←Employ robotic tools to support acquisition of samples, including soil, pebbles, hand-sized rock samples, and drill cores, manufactured in accordance with science requirements to minimize sample contamination		
F-74	Robot with arm capable of digging trenches, collecting samples of regolith, and holding instruments	UC-52	Utilize robots to document, collect, and stockpile samples on the lunar surface	←Employ sample containers appropriate for the specimens collected and science needs (e.g. contamination considerations, including bags, and sealed containers, that are accessible to robotic manipulation		

*Mapped to HLR Segment

ID	Functions	ID	Use Cases	← Characteristics & Needs	← Objectives	ID
F-13	Transport crew and systems from cislunar space to lunar surface South Pole sites	UC-21	Crewed missions to distributed landing sites around South Pole	←Emplace and operate science instrumentation in lunar and heliocentric orbits relevant to addressing the associated science objectives ←Emplace and operate science instrumentation on the lunar surface at locations relevant to addressing associated science objectives, including polar and non-polar locations on the lunar near side and far side	Enable long-term, planet-wide research by delivering science instruments to multiple science-relevant orbits and surface locations at the Moon and Mars.	SE-6
F-44	Transport crew and systems from cislunar space to lunar surface non-polar sites	UC-29	Crewed missions to non-polar landing sites			
F-18	Transport cargo from Earth to lunar surface	UC-28	Crew emplacement and set-up of science packages on lunar surface w/ long-term remote operation			
F-28	Conduct crew surface EVA activities					
F-51	Utilize robots to take measurements and conduct experiments on the surface					
F-33	Transport cargo from Earth to elements in deep space	UC-53	Crew emplacement and set-up of physics packages at cislunar elements w/ long-term remote operation			
F-58	Deliver free-flyers to cislunar space	UC-32	Autonomous deployment and long-term operation of free-flying packages in various lunar orbits			
N/A	N/A	UC-54	Develop and implement standards to govern communication systems operations to protect lunar far side environment	←Preserve radio free environment on far side ←Limit contamination of PSRs ←Protect sites of historic significance	Preserve and protect representative features of special interest, including lunar permanently shadowed regions and the radio quiet far side as well as Martian recurring slope lineae, to enable future high-priority science investigations.	SE-7
N/A	N/A	UC-55	Develop and implement standards to govern operations to protect PSR environment			
N/A	N/A	UC-56	Implement and comply with existing standards to protect sites of historic significance			

A.1.6 Applied Science

*Mapped to HLR Segment

ID	Functions	ID	Use Cases	← Characteristics & Needs	← Objectives	ID
F-18	Transport cargo from Earth to lunar surface	UC-28	Crew emplacement and set-up of science packages on lunar surface w/ long-term remote operation	←Emplace and operate science instrumentation in lunar and heliocentric orbits relevant to addressing the associated science objectives		AS-1
F-28	Conduct crew surface EVA activities					
F-51	Utilize robots to take measurements and conduct experiments on the surface					
F-18	Transport cargo from Earth to lunar surface	UC-57				

*Mapped to HLR Segment

ID	Functions	ID	Use Cases	← Characteristics & Needs	← Objectives	ID
F-28	Conduct crew surface EVA activities		Crew and/or robotic emplacement and set-up of science instrumentation in lunar orbit w/ long-term remote operation	←Emplace and operate science instrumentation on the lunar surface at locations relevant to addressing associated science objectives, including polar and non-polar locations on the lunar near side and far side	Characterize and monitor the contemporary environments of the lunar and Martian surfaces and orbits, including investigations of micrometeorite flux, atmospheric weather, space weather, space weathering, and dust, to plan, support, and monitor safety of crewed operations in these locations.	
F-58	Deliver free-flyers to cislunar space	UC-58	Autonomous deployment and long-term operation of free-flying packages in various lunar and heliocentric orbits			
F-75	Science and exploration instrumentation assets in a polar low-lunar orbit			←Emplace and operate science instrumentation in lunar and heliocentric orbits relevant to addressing the associated science objectives	Coordinate on-going and future science measurements from orbital and surface platforms to optimize human-led science campaigns on the Moon and Mars.	AS-2
F-71	Provide high bandwidth, high availability comms between cislunar space and Earth	UC-59	Use imagery and data collected from orbit to identify exploration sites of interest to NASA, contribute to real-time mission planning and monitoring, and evaluate exploration sites post-mission	←Emplace and operate science instrumentation on the lunar surface at locations relevant to addressing associated science objectives, including polar and non-polar locations on the lunar near side and far side		
F-58	Deliver free-flyers to cislunar space					
F-13	Transport crew and systems from cislunar space to lunar surface South Pole sites	UC-21	Crewed missions to distributed landing sites around South Pole	←Visit geologically diverse sites around the South Pole	Characterize accessible lunar and Martian resources, gather scientific research data, and analyze potential reserves to satisfy science and technology objectives and enable In-Situ Resource Utilization (ISRU) on successive missions.	AS-3
F-40	Provide robotic systems on lunar surface controlled from Earth and/or cislunar space	UC-23	Robotic survey of potential crewed landing sites to identify locations of interest	←Collect samples from multiple locations, including frozen samples from PSRs, on the lunar surface		
F-45	Provide robotic systems in PSRs on lunar surface controlled from Earth and/or cislunar space	UC-48	Robotic survey of PSRs near potential crewed landing sites to identify locations of interest	←Return to Earth a variety of types of samples collected from the lunar surface, including of soil, pebbles, and rocks		
F-30	Provide local unpressurized crew surface mobility			←Return to Earth a variety of samples from the lunar subsurface at a variety of depths		
F-31	Provide pressurized crew surface mobility	UC-24	Crew excursions to locations distributed around landing site	←Return to Earth frozen samples from a variety of depths in their pristine state		
F-24	Provide PNT capability on the lunar surface					
F-28	Conduct crew surface EVA activities			←Return select samples of soil, rock, subsurface materials in containers sealed in lunar vacuum		
F-46	Crew survey of areas of interest and identification of samples	UC-25	Crew EVA exploration and identification of samples			

*Mapped to HLR Segment

ID	Functions	ID	Use Cases	← Characteristics & Needs	← Objectives	ID
F-23	Provide high bandwidth, high availability comms between lunar surface and Earth			←Mobility to conduct prospecting traverses with appropriate scientific instrumentation and drill capabilities over sites of interest		
F-24	Provide PNT capability on the lunar surface			←Deploy scientific payloads at distances outside the blast zone of the ascent vehicle		
F-28	Conduct crew surface EVA activities			←Emplace and operate science instrumentation in lunar and heliocentric orbits relevant to addressing the associated science objectives		
F-46	Crew survey of areas of interest and identification of samples					
F-32	Recover and package surface samples	UC-26	Crew collection of samples from lighted areas			
F-47	Provide tools and containers to recover and package surface samples					
F-48	Store collected samples on lunar surface					
F-52	Conduct crew surface EVA activities into PSRs					
F-53	Crew and/or robotic survey of areas of interest and identification of samples in PSRs	UC-30	Crew/robotic collection of samples from PSRs			
F-54	Recover and package surface samples in PSRs					
F-55	Store collected samples on lunar surface at ambient/native temperature					
F-49	Transport cargo from lunar surface to Earth at ambient/native temperature	UC-97	Return of collected samples to Earth at ambient/native temperatures in sealed sample containers			
F-121	Recover conditioned samples at ambient/native temperature after splashdown					
F-42	Transport cargo from lunar surface to Earth	UC-20	Return of collected samples to Earth in sealed sample containers			
F-43	Recover samples after splashdown					
F-76	Demonstrate operation of bio-regenerative ECLSS	UC-60	Include demonstrations of bio-regenerative oxygen and water recovery sub-systems into ECLSS architecture in cislunar elements	←Demonstrate operation of bioregenerative ECLSS sub-systems in deep space.	Conduct applied scientific investigations essential for the development of bioregenerative-based, ecological life support systems	AS-4

*Mapped to HLR Segment

ID	Functions	ID	Use Cases	← Characteristics & Needs	← Objectives	ID
F-77	Demonstrate operation of plant growth system	UC-61	Include demonstrations of plant growth sub-systems into ECLSS architecture in cislunar elements	←Demonstrate operation of plant based ECLSS sub-systems in deep space.	Define crop plant species, including methods for their productive growth, capable of providing sustainable and nutritious food sources for lunar, Deep Space transit, and Mars habitation.	AS-5
F-33	Transport cargo from Earth to elements in deep space	UC-62	Crew emplacement and set-up of physics packages at cislunar elements w/ long-term remote operation	←Emplace and operate science packages in lunar orbit	Advance understanding of how physical systems and fundamental physical phenomena are affected by partial gravity, microgravity, and general environment of the Moon, Mars, and deep space transit.	AS-6
F-18	Transport cargo from Earth to lunar surface	UC-63	Crew emplacement and set-up of physics packages on lunar surface w/ long-term remote operation	←Emplace and operate science packages on the lunar surface		
F-28	Conduct crew surface EVA activities					
F-60	Provide IVA laboratory space on lunar surface	UC-43	Crew IVA research in dedicated science laboratory on the lunar surface	←Provide IVA laboratory space on the lunar surface and provide for crew time to conduct experiments		

A.2 INFRASTRUCTURE

*Mapped to HLR Segment

ID	Functions	ID	Use Cases	← Characteristics & Needs	← Objectives	ID
F-78	Provide power generation on lunar surface at crewed exploration sites	UC-64	Deployment of power generation and storage systems at multiple locations around the South Pole	←Emplace power generation and power storage capabilities on the lunar surface	Develop an incremental lunar power generation and distribution system that is evolvable to support continuous robotic/human operation and is capable of scaling to global power utilization and industrial power levels.	LI-1
F-79	Provide power storage on lunar surface at crewed exploration sites			←Emplace power distribution and storage capabilities on the lunar surface to allow power utilization at multiple locations around exploration sites		
F-80	Provide power distribution around generation/storage facilities on lunar surface	UC-65	Deployment of power distribution capabilities around generation and storage facilities on the surface			
F-23	Provide high bandwidth, high availability comms between lunar surface and Earth	UC-66	Deployment of assets in lunar orbit to provide high availability, high bandwidth communication from a variety of exploration locations on the lunar surface to Earth	←Implement communications systems to enable high bandwidth, high availability communications between Earth-based personnel, surface crew, and science packages on the surface	Develop a lunar surface, orbital, and Moon-to-Earth communications architecture capable of scaling to support long term science, exploration, and industrial needs.	LI-2
F-81	Provide Earth ground stations for exploration communications					
F-82	Provide high bandwidth, high availability comms between lunar surface assets	UC-67	Communications between assets on the lunar surface			

*Mapped to HLR Segment

ID	Functions	ID	Use Cases	← Characteristics & Needs	← Objectives	ID
F-24	Provide PNT capability on the lunar surface	UC-68	Deployment of assets in lunar orbit to provide high availability position, navigation, and timing for astronauts and robotic elements at exploration locations on the lunar surface	←Implement navigation and timing systems to enable high availability navigation on the surface	Develop a lunar position, navigation and timing architecture capable of scaling to support long term science, exploration, and industrial needs.	LI-3
F-83	Demonstrate capability for autonomous berm building	UC-69	Autonomous berm building or blast shield production building at crewed landing sites	←Deliver and demonstrate autonomous construction demonstration package(s) to the lunar South Pole	Demonstrate advanced manufacturing and autonomous construction capabilities in support of continuous human lunar presence and a robust lunar economy.	LI-4
F-84	Demonstrate capability for blast shield production					
F-85	Demonstrate capability for autonomous compaction of pathways	UC-70	Autonomous road building/route compaction and maintenance			
F-86	Demonstrate capability for maintenance of pathways					
F-87	Provide precision landing system for crew transport to lunar surface	UC-71	Landing of crew vehicle at specific pre-defined location within exploration area	←Demonstrate ability of lunar landers to reliably land within a defined radius around an intended location.	Demonstrate precision landing capabilities in support of continuous human lunar presence and a robust lunar economy.	LI-5
F-88	Provide precision landing system for cargo transport to lunar surface	UC-72	Landing of robotic landers at specific pre-defined location within exploration area			
F-30	Provide local unpressurized crew surface mobility	UC-73	Crew exploration around landing site or around habitation elements in EVA suits	←Demonstrate the ability to allow crew to move locally around landing sites to visit multiple locations of interest ←Demonstrate the ability to relocate surface elements to locations around the lunar South Pole between crewed surface missions.	Demonstrate local, regional, and global surface transportation and mobility capabilities in support of continuous human lunar presence and a robust lunar economy.	LI-6
F-24	Provide PNT capability on the lunar surface					
F-31	Provide pressurized crew surface mobility	UC-74	Crew relocation and exploration in a shirtsleeve environment			
F-24	Provide PNT capability on the lunar surface					
F-23	Provide high bandwidth, high availability comms between lunar surface and Earth	UC-75	Uncrewed relocation of mobility elements to landing sites around the South Pole			
F-24	Provide PNT capability on the lunar surface					
F-21	Generate power on lunar surface					
F-89	Semi-autonomous driving of mobility systems on surface					
F-90	Demonstrate operation of oxygen production ISRU	UC-76	Deploy and demonstrate operation of capability to recover oxygen from lunar regolith	←Deliver and demonstrate ISRU demonstration package(s) to the lunar South Pole		LI-7
F-91	Transport autonomous payloads from Earth to lunar surface					

*Mapped to HLR Segment

ID	Functions	ID	Use Cases	← Characteristics & Needs	← Objectives	ID
F-92	Unload autonomous payloads on lunar surface				Demonstrate industrial scale ISRU capabilities in support of continuous human lunar presence and a robust lunar economy.	
F-93	Demonstrate operation of water production ISRU					
F-91	Transport autonomous payloads from Earth to lunar surface	UC-77	Deploy and demonstrate operation of capability to recover polar water/volatiles			
F-92	Unload autonomous payloads on lunar surface					
F-12	Docking/berthing of spacecraft elements	UC-03	Aggregation and physical assembly of spacecraft elements in cislunar space	←Demonstrate capability to transfer propellant from one spacecraft to another in space	Demonstrate technologies supporting cislunar orbital/surface depots, construction and manufacturing maximizing the use of in-situ resources, and support systems needed for continuous human/robotic presence.	LI-8
F-94	Provide in space fueling of spacecraft	UC-78	Refueling of spacecraft in space	←Demonstrate capability to store propellant for extended durations in space		
F-95	Store cryogenic propellant in space	UC-79	Store cryogenic propellent for long durations in space			
F-96	Store cryogenic propellant on lunar surface	UC-80	Store cryogenic propellent for long durations on surface	←Demonstrate capability to store propellant on the lunar surface for extended durations		
F-56	Deliver utilization cargo to cislunar elements	UC-81	Deployment of assets to monitor the Sun in cislunar space	←Emplace systems to monitor solar weather and to predict SPEs	Develop environmental monitoring, situational awareness, and early warning capabilities to support a resilient, continuous human/robotic lunar presence.	LI-9
F-97	Provide solar observation and monitoring capability					

A.3 TRANSPORTATION & HABITATION

*Mapped to HLR Segment

ID	Functions	ID	Use Cases	← Characteristics & Needs	← Objectives	ID
F-01	Provide ground services				Develop cislunar systems that crew can routinely operate to and from lunar orbit and the lunar surface for extended durations.	TH-1
F-02	Stack and integrate	UC-01	Transport crew and systems from Earth to cislunar space	← Demonstrate transportation of crew and systems from Earth to stable lunar orbit		
F-03	Manage consumables and propellant					
F-04	Enable vehicle launch					

*Mapped to HLR Segment

ID	Functions	ID	Use Cases	← Characteristics & Needs	← Objectives	ID
F-05	Provide multiple launch attempts			← Demonstrate staged operation of crew transportation from stable lunar orbit with accessibility to both Earth and the lunar South Pole		
F-06	Provide aborts			← Demonstrate crew transport from stable lunar orbit to lunar surface and from lunar surface to NRHO		
F-07	Transport crew and systems from Earth to cislunar space					
F-08	Vehicle rendezvous, proximity ops, docking, and undocking in cislunar space	UC-02	Staging of crewed lunar surface missions from cislunar space	← Operate crew transportation system in uncrewed mode for extended periods on the lunar surface		
F-09	Provide PNT capability in cislunar space			← Demonstrate safe return to Earth of crew and systems from stable lunar orbit		
F-10	Provide crew habitation in cislunar space					
F-11	Transport elements from Earth to cislunar space	UC-03	Aggregation and physical assembly of spacecraft elements in cislunar space			
F-12	Docking/berthing of spacecraft elements					
F-13	Transport crew and systems from cislunar space to lunar surface South Pole sites	UC-04	Crew transport between cislunar space and lunar surface			
F-14	Transport crew and systems from lunar surface to cislunar space					
F-15	Operate crew vehicle in uncrewed mode on surface	UC-05	Human lander operates in standby mode while crew live in surface systems			
F-16	Transport crew and systems from cislunar space to Earth	UC-06	Return crew and systems from cislunar space to Earth			
F-17	Recover crew, systems, and cargo after splashdown					
F-18	Transport cargo from Earth to lunar surface	UC-07	Delivery of large elements and unloading of elements on lunar surface	← Demonstrate capabilities to deliver elements from Earth to the lunar surface	Develop system(s) that can routinely deliver a range of elements to the lunar surface.	TH-2
F-19	Unload cargo on lunar surface	UC-08	Deploy elements on the lunar surface	← Demonstrate unloading of cargo from delivery system(s)		
F-20	Reposition cargo on lunar surface					
F-21	Generate power on lunar surface			← Demonstrate capabilities to allow crew to live on the surface of the moon	Develop system(s) to allow crew to explore, operate, and live on the lunar surface and in lunar orbit with scalability to continuous presence; conducting scientific and industrial utilization as well as Mars analog activities.	TH-3
F-22	Store power on lunar surface					
F-23	Provide high bandwidth, high availability comms between lunar surface and Earth	UC-09	Crew operations on lunar surface	← Demonstrate capabilities to allow crew to exit habitable space and conduct EVA activities		
F-24	Provide PNT capability on the lunar surface					

*Mapped to HLR Segment

ID	Functions	ID	Use Cases	← Characteristics & Needs	← Objectives	ID
F-25	Provide pressurized, habitable environment on lunar surface	UC-10	Crew habitation in habitable elements on surface	← Demonstrate capabilities to allow crew to conduct science and utilization activities		
F-37	Remove trash from habitable elements on lunar surface					
F-18	Transport cargo from Earth to lunar surface					
F-27	Move cargo into habitable elements on lunar surface					
F-28	Conduct crew surface EVA activities	UC-11	Frequent crew EVA on surface			
F-29	Allow crew ingress/egress from habitable elements to vacuum					
F-30	Provide local unpressurized crew surface mobility	UC-12	Crew conduct utilization activities on surface			
F-31	Provide pressurized crew surface mobility					
F-32	Recover and package surface samples					
F-33	Transport cargo from Earth to elements in deep space	UC-13	Crew conduct utilization activities in cislunar space			
F-28	Conduct crew surface EVA activities	UC-14	Crew emplacement and set-up of science/utilization packages on lunar surface			
F-18	Transport cargo from Earth to lunar surface					
F-34	Provide pressurized, habitable environment in deep space	UC-15	Crew conduct long-duration increments in cislunar space	← Demonstrate in-space assembly of spacecraft in a stable lunar orbit with minimal orbital maintenance and support for power generation	Develop in-space and surface habitation system(s) for crew to live in deep space for extended durations, enabling future missions to Mars.	TH-4
F-33	Transport cargo from Earth to elements in deep space					
F-35	Move cargo into habitable elements in deep space					
F-36	Remove trash from habitable elements in deep space			← Demonstrate capabilities to allow crew to live in deep space for extended durations		
F-25	Provide pressurized, habitable environment on lunar surface	UC-10	Crew habitation in habitable elements on surface			
F-18	Transport cargo from Earth to lunar surface					
F-27	Move cargo into habitable elements on lunar surface					
F-37	Remove trash from habitable elements on lunar surface					
F-38	Provide crew health maintenance	UC-16	Crew emergency health care and monitoring while in transit	← Demonstrate remote crew health sub-system(s) in NRHO		TH-

*Mapped to HLR Segment

ID	Functions	ID	Use Cases	← Characteristics & Needs	← Objectives	ID
F-39	Provide crew remote medical systems in cislunar space	UC-17	Remote diagnosis and treatment of crew health during extended increments in cislunar space	← Demonstrate remote crew health sub-system(s) on the lunar surface	Develop systems that monitor and maintain crew health and performance throughout all mission phases, including during communication delays to Earth, and in an environment that does not allow emergency evacuation or terrestrial medical assistance.	
F-38	Provide crew health maintenance					
F-26	Provide crew remote medical systems on lunar surface	UC-18	Remote diagnosis and treatment of crew health during extended increments on lunar surface			
F-38	Provide crew health maintenance					
F-40	Provide robotic systems on lunar surface controlled from Earth and/or cislunar space	UC-19	Robotic assistance of crew exploration, surveying sites, locating samples and resources, and retrieval of samples	← Conduct operations in which robotic sub-systems, controlled remotely from Earth, NRHO, or the surface, support crew exploration	Develop integrated human and robotic systems with inter-relationships that enable maximum science and exploration during lunar missions.	TH-9
F-41	Provide robotic systems in cislunar space controlled from Earth and/or cislunar space					
F-42	Transport cargo from lunar surface to Earth	UC-20	Return of collected samples to Earth in sealed sample containers	← Demonstrate capabilities to return cargo from the lunar surface back to Earth	Develop systems capable of returning a range of cargo mass from the lunar surface to Earth, including the capabilities necessary to meet scientific and utilization objectives.	TH-11
F-43	Recover samples after splashdown					

A.4 OPERATIONS

*Mapped to HLR Segment

ID	Functions	ID	Use Cases	← Characteristics & Needs	← Objectives	ID
F-34	Provide pressurized, habitable environment in deep space	UC-35	Analog missions with extended durations in NRHO, followed by lunar surface missions	←Conduct mid-duration crew exploration on the lunar surface ←Conduct long-duration crew exploration in cis lunar space ←Conduct crewed and uncrewed testing of in-space habitation systems	Conduct human research and technology demonstrations on the surface of Earth, low Earth orbit platforms, cislunar platforms, and on the surface of the moon, to evaluate the effects of extended mission durations on the performance of crew and systems, reduce risk, and shorten the timeframe for system testing and readiness prior to the initial human Mars exploration campaign.	OP-1
F-33	Transport cargo from Earth to elements in deep space					
F-35	Move cargo into habitable elements in deep space					
F-98	Provide crew remote medical systems in deep space					
F-99	Transport crew and systems from cislunar space to lunar surface					

*Mapped to HLR Segment

ID	Functions	ID	Use Cases	← Characteristics & Needs	← Objectives	ID
F-100	Autonomous element command control in cislunar space	UC-82	Conduct autonomous/semi-autonomous mission operations in cislunar space	←Provide onboard autonomy to train, plan, and execute a safe mission ←Provide flight control and mission integration to ensure safety and mission success	Optimize operations, training and interaction between the team on Earth, crew members on orbit, and a Martian surface team considering communication delays, autonomy level, and time required for an early return to the Earth.	OP-2
F-59	Autonomous crew landing on lunar surface	UC-83	Conduct autonomous/semi-autonomous mission operations on lunar surface			
F-101	Autonomous element command control on lunar surface					
F-13	Transport crew and systems from cislunar space to lunar surface South Pole sites	UC-21	Crewed missions to distributed landing sites around South Pole	←Visit geographically diverse sites around the South Pole and non-polar regions ←Emplace science packages at distributed sites on the lunar surface ←Identify and collect samples from multiple locations, including frozen samples from PSRs, on the lunar surface ←Return samples collected on the surface, including frozen samples, to Earth	Characterize accessible resources, gather scientific research data, and analyze potential reserves to satisfy science and technology objectives and enable use of resources on successive missions.	OP-3
F-40	Provide robotic systems on lunar surface controlled from Earth and/or cislunar space	UC-23	Robotic survey of potential crewed landing sites to identify locations of interest			
F-45	Provide robotic systems in PSRs on lunar surface controlled from Earth and/or cislunar space	UC-48	Robotic survey of PSRs near potential crewed landing sites to identify locations of interest			
F-30	Provide local unpressurized crew surface mobility	UC-24	Crew excursions to locations distributed around landing site			
F-31	Provide pressurized crew surface mobility					
F-24	Provide PNT capability on the lunar surface					
F-28	Conduct crew surface EVA activities	UC-25	Crew EVA exploration and identification of samples			
F-46	Crew survey of areas of interest and identification of samples					
F-23	Provide high bandwidth, high availability comms between lunar surface and Earth					
F-24	Provide PNT capability on the lunar surface					
F-28	Conduct crew surface EVA activities	UC-26	Crew collection of samples from lighted areas			
F-46	Crew survey of areas of interest and identification of samples					
F-32	Recover and package surface samples					
F-47	Provide tools and containers to recover and package surface samples					
F-48	Store collected samples on lunar surface					
F-52	Conduct crew surface EVA activities into PSRs	UC-30	Crew/robotic collection of samples from PSRs			

*Mapped to HLR Segment

ID	Functions	ID	Use Cases	← Characteristics & Needs	← Objectives	ID
F-53	Crew and/or robotic survey of areas of interest and identification of samples in PSRs					
F-54	Recover and package surface samples in PSRs					
F-55	Store collected samples on lunar surface at ambient/native temperature					
F-49	Transport cargo from lunar surface to Earth at ambient/native temperature	UC-97	Return of collected samples to Earth at ambient/native temperatures in sealed sample containers			
F-121	Recover conditioned samples at ambient/native temperature after splashdown					
F-42	Transport cargo from lunar surface to Earth	UC-20	Return of collected samples to Earth in sealed sample containers			
F-43	Recover samples after splashdown					
F-80	Provide power distribution around generation/storage facilities on lunar surface	UC-84	Utilize common interface for power and commodity transfers on the surface	← Integrate networks and mission systems to exchange data between Earth and campaign systems	Establish command control processes, common interfaces, and ground systems that will support expanding human missions at the Moon and Mars.	OP-4
F-102	Provide common power distribution interface					
F-103	Provide common interface for water and gas transfer on lunar surface	UC-85	Utilize common interface for tool integration to elements			
F-104	Provide robotic manipulator interface on lunar surface					
F-28	Conduct crew surface EVA activities	UC-86	Conduct crew EVA exploration on the lunar surface	← Transport crew and cargo between landing or base site and exploration sites of varying distances ← Conduct EVA activities utilizing mobility assets and tools	Operate surface mobility systems, e.g., extra-vehicular activity (EVA) suits, tools and vehicles.	OP-5
F-29	Allow crew ingress/egress from habitable elements to vacuum					
F-23	Provide high bandwidth, high availability comms between lunar surface and Earth					
F-30	Provide local unpressurized crew surface mobility	UC-87	Crew driving of mobility systems in EVA suits			
F-31	Provide pressurized crew surface mobility	UC-88	Crew operation of mobility systems in shirt sleeve environment			
F-24	Provide PNT capability on the lunar surface					
F-105	Provide EVA tools to collect samples	UC-89	Crew use of EVA tools to collect samples, clean suits, etc.			
F-106	Provide EVA tools to clean EVA suits and equipment					

*Mapped to HLR Segment

ID	Functions	ID	Use Cases	← Characteristics & Needs	← Objectives	ID
F-34	Provide pressurized, habitable environment in deep space	UC-35	Analog missions with extended durations in NRHO, followed by lunar surface missions	←Conduct mid-duration crew exploration on the lunar surface ←Conduct long-duration crew exploration in cis lunar space ←Transition crew from micro-gravity to partial gravity	Evaluate, understand, and mitigate the impacts on crew health and performance of a long deep space orbital mission, followed by partial gravity surface operations on the Moon.	OP-6
F-33	Transport cargo from Earth to elements in deep space					
F-35	Move cargo into habitable elements in deep space					
F-39	Provide crew remote medical systems in cislunar space					
F-59	Autonomous crew landing on lunar surface					
F-25	Provide pressurized, habitable environment on lunar surface					
F-26	Provide crew remote medical systems on lunar surface					
F-18	Transport cargo from Earth to lunar surface					
F-27	Move cargo into habitable elements on lunar surface					
F-38	Provide crew health maintenance	UC-16	Crew emergency health care and monitoring while in transit	←Conduct extended crewed and uncrewed testing of in-space habitation system ←Provide crew health and performance capabilities for Mars duration mission ←Provide crew survival capabilities	Validate readiness of systems and operations to support crew health and performance for the initial human Mars exploration campaign.	OP-7
F-39	Provide crew remote medical systems in cislunar space	UC-17	Remote diagnosis and treatment of crew health during extended increments in cislunar space			
F-38	Provide crew health maintenance					
F-26	Provide crew remote medical systems on lunar surface	UC-18	Remote diagnosis and treatment of crew health during extended increments on lunar surface			
F-38	Provide crew health maintenance					
F-107	Transport crew to lunar surface in proximity of previous exploration campaign landing site	UC-90	Crew recovery of excess propellant from tanks of previous landers	←Access and re-use surface assets from previous crewed and uncrewed missions	Demonstrate the capability to find, service, upgrade, or utilize instruments and equipment from robotic landers or previous human missions on the surface of the Moon and Mars.	OP-8
F-108	Recover propellant residuals from lander tanks					
F-109	Store propellant on lunar surface					
F-110	Transport crew to lunar surface in proximity of mobility assets on lunar surface	UC-91	Crew reuse of mobility assets on surface			
F-111	Operate mobility asset in dormancy/remote mode between crewed missions					
F-112	Transport crew to lunar surface in proximity of habitation assets on lunar surface	UC-92	Crew reuse of habitation assets on surface			

ID	Functions	ID	Use Cases	← Characteristics & Needs	← Objectives	ID
F-113	Operate habitation asset in dormancy/remote mode between crewed missions					
F-114	Provide robotic systems on lunar surface controlled by crew	UC-93	Robotic assistance of crew exploration, surveying sites, locating samples and resources, and retrieval of samples, controlled by surface crew	←Provide autonomous and remote operations of surface systems from external systems, including Earth, orbital, and other surface locations ←Ensure safe interactions between crew and automated/autonomous systems	Demonstrate the capability of integrated robotic systems to support and maximize the useful work performed by crewmembers on the surface, and in orbit.	OP-9
F-40	Provide robotic systems on lunar surface controlled from Earth and/or cislunar space	UC-94	Robotic assistance of crew exploration, surveying sites, locating samples and resources, and retrieval of samples, controlled from Earth or cislunar space	←Provide autonomous and remote operations of surface systems from external systems, including Earth, orbital, and other surface locations ←Ensure safe interactions between crew and automated/autonomous systems	Demonstrate the capability to operate robotic systems that are used to support crew members on the lunar or Martian surface, autonomously or remotely from the Earth or from orbiting platforms.	OP-10
F-115	Produce ISRU oxygen on lunar surface	UC-95	Deploy capability to recover oxygen from lunar regolith, store processed oxygen, and transfer to habitation elements	←Emplace and operate ISRU demonstration packages on the lunar surface	Demonstrate the capability to use commodities produced from planetary surface or in-space resources to reduce the mass required to be transported from Earth.	OP-11
F-116	Store ISRU oxygen on lunar surface					
F-117	Transfer ISRU oxygen to exploration elements on lunar surface					
F-91	Transport autonomous payloads from Earth to lunar surface					
F-92	Unload autonomous payloads on lunar surface					
F-118	Produce ISRU water on lunar surface	UC-96	Deploy capability to recover polar water, store products, and transfer to habitation elements			
F-119	Store ISRU water on lunar surface					
F-120	Transfer ISRU water to exploration elements on lunar surface					
F-91	Transport autonomous payloads from Earth to lunar surface					
F-92	Unload autonomous payloads on lunar surface					

*Mapped to HLR Segment

ID	Functions	ID	Use Cases	← Characteristics & Needs	← Objectives	ID
N/A	N/A	N/A	N/A	←Preserve radio-free environment on far side ←Limit contamination of PSRs ←Demonstrate recovery of excess fluids and gases from lunar landers	Establish procedures and systems that will minimize the disturbance to the local environment, maximize the resources available to future explorers, and allow for reuse/recycling of material transported from Earth (and from the lunar surface in the case of Mars) to be used during exploration.	OP-12

A.5 LIST OF USE CASES

ID	Use Case
UC-01	Transport crew and systems from Earth to cislunar space
UC-02	Staging of crewed lunar surface missions from cislunar space
UC-03	Aggregation and physical assembly of spacecraft elements in cislunar space
UC-04	Crew transport between cislunar space and lunar surface
UC-05	Human lander operates in standby mode while crew live in surface systems
UC-06	Return crew and systems from cislunar space to Earth
UC-07	Delivery of large elements and unloading of elements on lunar surface
UC-08	Deploy elements on the lunar surface
UC-09	Crew operations on lunar surface
UC-10	Crew habitation in habitable elements on surface
UC-11	Frequent crew EVA on surface
UC-12	Crew conduct utilization activities on surface
UC-13	Crew conduct utilization activities in cislunar space
UC-14	Crew emplacement and set-up of science/utilization packages on lunar surface
UC-15	Crew conduct long-duration increments in cislunar space
UC-16	Crew emergency health care and monitoring while in transit
UC-17	Remote diagnosis and treatment of crew health during extended increments in cislunar space
UC-18	Remote diagnosis and treatment of crew health during extended increments on lunar surface
UC-19	Robotic assistance of crew exploration, surveying sites, locating samples and resources, and retrieval of samples
UC-20	Return of collected samples to Earth in sealed sample containers
UC-21	Crewed missions to distributed landing sites around South Pole
UC-22	Crewed missions to non-polar landing sites, including the lunar far side
UC-23	Robotic survey of potential crewed landing sites to identify locations of interest
UC-24	Crew excursions to locations distributed around landing site
UC-25	Crew EVA exploration and identification of samples
UC-26	Crew collection of samples from lighted areas
UC-27	Orbital surveys before, during, and after crew missions
UC-28	Crew emplacement and set-up of science packages on lunar surface w/ long-term remote operation
UC-29	Crewed missions to non-polar landing sites
UC-30	Crew/robotic collection of samples from PSRs
UC-31	Crew emplacement and set-up of Heliophysics packages at cislunar elements w/ long-term remote operation
UC-32	Autonomous deployment and long-term operation of free-flying packages in various lunar orbits
UC-33	Crew emplacement and set-up of Heliophysics packages on lunar surface w/ long-term remote operation
UC-34	Crew collection of regolith samples from a variety of sites

UC-35	Analog missions with extended durations in NRHO, followed by lunar surface missions
UC-36	IVA facilities (e.g., instruments, racks, stowage, power) on the lunar surface to enable biological science analyses
UC-37	IVA facilities (e.g., instruments, racks, stowage, power) in cislunar orbit to enable biological science analyses
UC-38	Analog missions with extended durations in NRHO, followed by lunar surface missions, with autonomous/semi-autonomous operations
UC-39	Crew emplacement and set-up of astrophysics instrumentation on far side lunar surface w/ long-term remote operation
UC-40	Crew conduct fundamental physics experiments while in the habitable volume on the lunar surface
UC-41	Crew emplacement and set-up of fundamental physics experiments at cislunar elements w/ long-term remote operation
UC-42	Crew emplacement and set-up of fundamental physics experiments on lunar surface w/ long-term remote operation
UC-43	Crew IVA research in dedicated science laboratory on the lunar surface
UC-44	Crew stows samples collected during fundamental physics experiments for return to Earth
UC-45	Provide advanced geology training as well as detailed objective-specific training to astronauts for science activities prior to each Artemis mission
UC-46	Train astronauts for science tasks during an Artemis mission utilizing in situ training capabilities
UC-47	Allow ground personnel and science team to directly engage with astronauts on the surface and in lunar orbit, augmenting the crew's effectiveness at conducting science activities.
UC-48	Robotic survey of PSRs near potential crewed landing sites to identify locations of interest
UC-49	Crew excursions to PSRs near landing site
UC-50	Utilize robots to survey sites around the South Pole.
UC-51	Utilize robots to take measurements and conduct experiments on the surface
UC-52	Utilize robots to document, collect, and stockpile samples on the lunar surface
UC-53	Crew emplacement and set-up of physics packages at cislunar elements w/ long-term remote operation
UC-54	Develop and implement standards to govern communication systems operations to protect lunar far side environment
UC-55	Develop and implement standards to govern operations to protect PSR environment
UC-56	Implement and comply with existing standards to protect sites of historic significance
UC-57	Crew and/or robotic emplacement and set-up of science instrumentation in lunar orbit w/ long-term remote operation
UC-58	Autonomous deployment and long-term operation of free-flying packages in various lunar and heliocentric orbits
UC-59	Use imagery and data collected from orbit to identify exploration sites of interest to NASA, contribute to real-time mission planning and monitoring, and evaluate exploration sites post-mission
UC-60	Include demonstrations of bio-regenerative oxygen and water recovery sub-systems into ECLSS architecture in cislunar elements
UC-61	Include demonstrations of plant growth sub-systems into ECLSS architecture in cislunar elements
UC-62	Crew emplacement and set-up of physics packages at cislunar elements with long-term remote operation
UC-63	Crew emplacement and set-up of physics packages on lunar surface with long-term remote operation

UC-64	Deployment of power generation and storage systems at multiple locations around the South Pole
UC-65	Deployment of Power distribution capabilities around generation and storage facilities on the surface
UC-66	Deployment of assets in lunar orbit to provide high availability, high bandwidth communication from a variety of exploration locations on the lunar surface to Earth
UC-67	Communications between assets on the lunar surface
UC-68	Deployment of assets in lunar orbit to provide high availability position, navigation, and timing for astronauts and robotic elements at exploration locations on the lunar surface
UC-69	Autonomous berm building or blast shield production building at crewed landing sites
UC-70	Autonomous road building/route compaction and maintenance
UC-71	Landing of crew vehicle at specific pre-defined location within exploration area
UC-72	Landing of robotic landers at specific pre-defined location within exploration area
UC-73	Crew exploration around landing site or around habitation elements in EVA suits
UC-74	Crew relocation and exploration in a shirtsleeve environment
UC-75	Uncrewed relocation of mobility elements to landing sites around the South Pole
UC-76	Deploy and demonstrate operation of capability to recover oxygen from lunar regolith
UC-77	Deploy and demonstrate operation of capability to recover polar water/volatiles
UC-78	Refueling of spacecraft in space
UC-79	Store cryogenic propellent for long durations in space
UC-80	Store cryogenic propellent for long durations on surface
UC-81	Deployment of assets to monitor the Sun in cislunar space
UC-82	Conduct autonomous/semi-autonomous mission operations in cislunar space
UC-83	Conduct autonomous/semi-autonomous mission operations on lunar surface
UC-84	Utilize common interface for power and commodity transfers on the surface
UC-85	Utilize common interface for tool integration to elements
UC-86	Conduct crew EVA exploration on the lunar surface
UC-87	Crew driving of mobility systems in EVA suits
UC-88	Crew operation of mobility systems in shirt sleeve environment
UC-89	Crew use of EVA tools to collect samples, clean suits, etc.
UC-90	Crew recovery of excess propellant from tanks of previous landers
UC-91	Crew reuse of mobility assets on surface
UC-92	Crew reuse of habitation assets on surface
UC-93	Robotic assistance of crew exploration, surveying sites, locating samples and resources, and retrieval of samples, controlled by surface crew
UC-94	Robotic assistance of crew exploration, surveying sites, locating samples and resources, and retrieval of samples, controlled from Earth or cislunar space
UC-95	Deploy capability to recover oxygen from lunar regolith, store processed oxygen, and transfer to habitation elements
UC-96	Deploy capability to recover polar water, store products, and transfer to habitation elements
UC-97	Return of collected samples to Earth at ambient/native temperatures in sealed sample containers

A.5 LIST OF FUNCTIONS

ID	Function
F-01	Provide ground services
F-02	Stack and integrate
F-03	Manage consumables and propellant
F-04	Enable vehicle launch
F-05	Provide multiple launch attempts
F-06	Provide aborts
F-07	Transport crew and systems from Earth to cislunar space
F-08	Vehicle rendezvous, proximity ops, docking, and undocking in cislunar space
F-09	Provide PNT capability in cislunar space
F-10	Provide crew habitation in cislunar space
F-11	Transport elements from Earth to cislunar space
F-12	Docking/berthing of spacecraft elements
F-13	Transport crew and systems from cislunar space to lunar surface South Pole sites
F-14	Transport crew and systems from lunar surface to cislunar space
F-15	Operate crew vehicle in uncrewed mode on surface
F-16	Transport crew and systems from cislunar space to Earth
F-17	Recover crew, systems, and cargo after splashdown
F-18	Transport cargo from Earth to lunar surface
F-19	Unload cargo on lunar surface
F-20	Reposition cargo on lunar surface
F-21	Generate power on lunar surface
F-22	Store power on lunar surface
F-23	Provide high bandwidth, high availability comms between lunar surface and Earth
F-24	Provide PNT capability on the lunar surface
F-25	Provide pressurized, habitable environment on lunar surface
F-26	Provide crew remote medical systems on lunar surface
F-27	Move cargo into habitable elements on lunar surface
F-28	Conduct crew surface EVA activities
F-29	Allow crew ingress/egress from habitable elements to vacuum
F-30	Provide local unpressurized crew surface mobility
F-31	Provide pressurized crew surface mobility
F-32	Recover and package surface samples
F-33	Transport cargo from Earth to elements in deep space
F-34	Provide pressurized, habitable environment in deep space
F-35	Move cargo into habitable elements in deep space
F-36	Remove trash from habitable elements in deep space

F-37	Remove trash from habitable elements on lunar surface
F-38	Provide crew health maintenance
F-39	Provide crew remote medical systems in cislunar space
F-40	Provide robotic systems on lunar surface controlled from Earth and/or cislunar space
F-41	Provide robotic systems in cislunar space controlled from Earth and/or cislunar space
F-42	Transport cargo from lunar surface to Earth
F-43	Recover samples after splashdown
F-44	Transport crew and systems from cislunar space to lunar surface non-polar sites
F-45	Provide robotic systems in PSRs on lunar surface controlled from Earth and/or cislunar space
F-46	Crew survey of areas of interest and identification of samples
F-47	Provide tools and containers to recover and package surface samples
F-48	Store collected samples on lunar surface
F-49	Transport cargo from lunar surface to Earth at ambient/native temperature
F-50	Orbital observation and sensing of lunar surface
F-51	Provide surface power for deployed payloads
F-52	Conduct crew surface EVA activities into PSRs
F-53	Crew and/or robotic survey of areas of interest and identification of samples in PSRs
F-54	Recover and package surface samples in PSRs
F-55	Store collected samples on lunar surface at ambient/native temperature
F-56	Deliver utilization cargo to cislunar elements
F-57	External mounting points on cislunar elements
F-58	Deliver free-flyers to cislunar space
F-59	Autonomous crew landing on lunar surface
F-60	Provide IVA laboratory space on lunar surface
F-61	Transport cargo from lunar surface to Earth at freezing temperatures
F-62	Provide IVA laboratory space in cislunar space
F-63	Aggregate and extended storage collected samples in cislunar space
F-64	Transport cargo from cislunar space to Earth at freezing temperatures
F-65	Transport crew and systems from cislunar space to lunar surface far side sites
F-66	Transport cargo to the far side lunar surface
F-67	Conduct crew far side surface EVA activities
F-68	Provide training of crew prior to mission
F-69	Provide in situ training of crew in cislunar space
F-70	Provide in situ training of crew on lunar surface
F-71	Provide high bandwidth, high availability comms between cislunar space and Earth
F-72	Provide local unpressurized crew surface mobility into PSRs
F-73	Provide robotic systems on lunar surface controlled from Earth and/or cislunar space b
F-74	Robot with arm capable of digging trenches, collecting samples of regolith, and holding instruments
F-75	Science and exploration instrumentation assets in a polar low-lunar orbit

F-76	Demonstrate operation of bio-regenerative ECLSS
F-77	Demonstrate operation of plant growth system
F-78	Provide power generation on lunar surface at crewed exploration sites
F-79	Provide power storage on lunar surface at crewed exploration sites
F-80	Provide power distribution around generation/storage facilities on lunar surface
F-81	Provide Earth ground stations for exploration communications
F-82	Provide high bandwidth, high availability comms between lunar surface assets
F-83	Demonstrate capability for autonomous berm building
F-84	Demonstrate capability for blast shield production
F-85	Demonstrate capability for autonomous compaction of pathways
F-86	Demonstrate capability for maintenance of pathways
F-87	Provide precision landing system for crew transport to lunar surface
F-88	Provide precision landing system for cargo transport to lunar surface
F-89	Semi-autonomous driving of mobility systems on surface
F-90	Demonstrate operation of oxygen production ISRU
F-91	Transport autonomous payloads from Earth to lunar surface
F-92	Unload autonomous payloads on lunar surface
F-93	Demonstrate operation of water production ISRU
F-94	Provide in space fueling of spacecraft
F-95	Store cryogenic propellant in space
F-96	Store cryogenic propellant on lunar surface
F-97	Provide solar observation and monitoring capability
F-98	Provide crew remote medical systems in deep space
F-99	Transport crew and systems from cislunar space to lunar surface
F-100	Autonomous element command control in cislunar space
F-101	Autonomous element command control on lunar surface
F-102	Provide common power distribution interface
F-103	Provide common interface for water and gas transfer on lunar surface
F-104	Provide robotic manipulator interface on lunar surface
F-105	Provide EVA tools to collect samples
F-106	Provide EVA tools to clean EVA suits and equipment
F-107	Transport crew to lunar surface in proximity of previous exploration campaign landing site
F-108	Recover propellant residuals from lander tanks
F-109	Store propellant on lunar surface
F-110	Transport crew to lunar surface in proximity of mobility assets on lunar surface
F-111	Operate mobility asset in dormancy/remote mode between crewed missions
F-112	Transport crew to lunar surface in proximity of habitation assets on lunar surface
F-113	Operate habitation asset in dormancy/remote mode between crewed missions
F-114	Provide robotic systems on lunar surface controlled by crew

F-115	Produce ISRU oxygen on lunar surface
F-116	Store ISRU oxygen on lunar surface
F-117	Transfer ISRU oxygen to exploration elements on lunar surface
F-118	Produce ISRU water on lunar surface
F-119	Store ISRU water on lunar surface
F-120	Transfer ISRU water to exploration elements on lunar surface
F-121	Recover conditioned samples at ambient/native temperature after splashdown

APPENDIX B: ACRONYMS, ABBREVIATIONS, AND GLOSSARY OF TERMS

B.1 ACRONYMS AND ABBREVIATIONS

ACR	Architecture Concept Review
ADD	Architecture Definition Document
AFS	Augmented Forward Signal
AS	Applied Sciences
ASA	Australian Space Agency
ASI	Italian Space Agency (Agenzia Spaziale Italiana)
AU	Astronomical Unit
BEO	Beyond Earth Orbit
CAPSTONE	Cislunar Autonomous Positioning System Technology Operations and Navigation Experiment
CLPS	Commercial Lunar Payload Services
CM	Crew Module
CNES	Centre National D'Etudes Spatiales
CPNT	Communication, Positioning, Navigation, and Timing
CMP	Co-Manifested Payload
CSA	Canadian Space Agency
DLR	German Aerospace Center
DSN	Deep Space Network
DSL	Deep Space Logistics
DST	Deep Space Transport
DTE	Direct-to-Earth
ECLS	Environmental Control and Life Support
EDL	Entry, Descent, and Landing
EDLA	Entry, Descent, Landing, and Ascent
EGS	Exploration Ground Systems
EP	Electric Propulsion
ESA	European Space Agency
ESM	European Service Module
EUS	Exploration Upper Stage
EVA	Extravehicular Activity
FE	Foundational Exploration
FSP	Fission Surface Power
GERS	Gateway Extravehicular Robotic System
GNSS	Global Navigation Satellite System

HALO	Habitation and Logistics Outpost
HBS	Human and Biological Science
HIAD	Hypersonic Inflatable Aerodynamic Decelerator
HLCS	HALO Lunar Communications Systems
HLR	Human Lunar Return
HLS	Human Landing System
HS	Heliophysics Science
iCPS	Interim Cryogenic Propulsion Stage
ICSIS	International Communication System Interoperability Standards
I-HAB	International Habitation Module
IMEWG	International Mars Exploration Working Group
IOAG	Interagency Operations Advisory Group
ISA	Israel Space Agency
ISECG	International Space Exploration Coordination Group
ISLSWG	International Space Life Sciences Working Group
ISRO	Indian Space Research Organization
ISRU	In-Situ Resource Utilization
ISS	International Space Station
IVA	Intra-Vehicular Activities
JAXA	Japan Aerospace Exploration Agency
KASI	Korea Astronomy and Space Science Institute
KPLO	Korea Pathfinder Lunar Orbiter
LAS	Launch Abort System
LCRNS	Lunar Communication Relay and Navigation System
LEAG	Lunar Exploration Analysis Group
LEGS	Lunar Exploration Ground System
LEO	Low-Earth Orbit
LI	Lunar Infrastructure
LNIS	LunaNet Interoperability Specification
LOC	Loss of Crew
LOFTID	Low-Earth Orbit Flight Test of an Inflatable Decelerator
LOM	Loss of Mission
LPS	Lunar/Planetary Science
M2M	Moon-to-Mars
MAV	Mars Ascent Vehicle
MDS	Mars Descent System
MEPAG	Mars Exploration Program Analysis Group
ML	Mobile Launcher
ML2	Mobile Launcher 2
NASA	National Aeronautics and Space Administration

NEP	Nuclear Electric Propulsion
NextSTEP	Next Space Technologies for Exploration Partnerships
NRHO	Near Rectilinear Halo Orbit
NTP	Nuclear Thermal Propulsion
OP	Operations
PPE	Power Propulsion Element
PPS	Physics and Physical Sciences
PRISM	Payload and Research Investigations from the Surface of the Moon
PSR	Permanently Shadowed Regions
SAC	Strategic Analysis Cycles
SE	Science-Enabling
SEP	Solar Electric Propulsion
SLE	Sustained Lunar Evolution
SLS	Space Launch System
SM	Service Module
SMD	Science Mission Directorate
SRTs	Safety Reporting Thresholds
SSERVI	Solar System Exploration Research Virtual Institute
TH	Transportation and Habitation
UHF	Ultra-High Frequency
VAB	Vehicle Assembly Building
VIPER	Volatiles Investigating Polar Exploration Rover
xEVA	Exploration Extra-Vehicular Activity
xEVAS	Exploration Extra-Vehicular Activity Services

B.2 GLOSSARY OF TERMS

Term	Description
Architecture	The high-level unifying structure that defines a system. It provides a set of rules, guidelines, and constraints that defines a cohesive and coherent structure consisting of constituent parts, relationships and connections that establish how those parts fit and work together. (Definition from NASA's System Engineering Handbook)
Artemis Mission	The crewed portion of an Artemis Mission Campaign, beginning at crew liftoff from Earth and ending at crew return to Earth.
Artemis Mission Campaign	A collective grouping of uncrewed missions and their associated crewed mission.
Automation	Automatically controlled operation of an apparatus, process, or system by mechanical or electronic devices that take the place of human labor (e.g. computer control of a docking operation or vehicle surface traverse). Human intervention can be available, as determined by hazard controls (e.g. breakout or transition to safe mode), but not required to complete an autonomous operation.
Autonomy	The ability of a system to achieve goals while operating independently of external control. Autonomy does not preclude external re-prioritization or generation of new goals. It only requires execution of existing goals without external control.
Baseline	An agreed-to set of requirements, designs, or documents that will have changes controlled through a formal approval and monitoring process.
Campaign	A series of interrelated missions that together achieve Agency goals and objectives. (Definition from M2M Objectives)
Characteristics	Features or activities of exploration mission implementation that are necessary to satisfy the Goals and Objectives.
Cislunar	The region of space from the Earth to the Moon. Specifically for the Moon-to-Mars Architecture, elements under the influence of lunar gravity
Co-Manifested Payload	Cargo on a transportation element utilizing excess volume and mass, e.g., cargo located inside the payload attach fitting adapter ring.
Concept of Operations	Developed early in Pre-Phase A, describes the overall high-level concept of how the system will be used to meet stakeholder expectations, usually in a time sequenced manner. It describes the system from an operational perspective and helps facilitate an understanding of the system goals. It stimulates the development of the requirements and architecture related to the user elements of the system. It serves as the basis for subsequent definition documents and provides the foundation for the long-range operational planning activities (for nominal and contingency operations). It provides the criteria for the validation of the system. In cases where an Operations Concept (OpsCon) is developed, the concept of operations feeds into the OpsCon and they evolve together. The concept of operations becomes part of the Concept Documentation.
Continuous Presence	Steady cadence of human/robotic missions in subject orbit/surface with the desired endpoint of 365/24/7 operations. (Definition from M2M Objectives)
Control Mass	Used to define the capability and baseline architecture of the system. It represents the controlled, not-to-exceed allocation of mass to an element.

Term	Description
Deep Space Environments	Deep Space is the vast region of space that extends to interplanetary space, to Mars and beyond. It is the region of space beyond the dark side of our Moon, including Lagrange 2, or L2, (274,000 miles from Earth). The environment in this vast region of space has many defining factors that includes harsh radiation (both solar particle events and galactic cosmic rays), space weather, and micro-gravity.
Deep Space Transport (DST)	The term DST is used to describe the assembled Mars transit vehicle stack, which will consist of a propulsion and power transportation system backbone and attached cargo. There are two DST variants: in the crew variant, the cargo will consist of a transit habitat that may or may not be a separate free-flyer that docks with transport; in the cargo variant, the cargo will consist of orbital assets to be delivered to Mars orbit, or surface assets mounted to Mars Descent Systems that will be delivered to the Mars surface.
Demonstrate	Deploy an initial capability to enable system maturation and future industry growth in alignment with architecture objectives. (Definition from M2M Objectives)
Develop	Design, build, and deploy a system, ready to be operated by the user, to fully meet architectural objectives. (Definition from M2M Objectives)
Effectivity	The conditions or mission for which a requirement is initially applicable.
Element	A notional exploration system enabling a high-level functional allocation, e.g. crew transport, habitation, logistics delivery, etc.
Explore	Excursion-based expeditions focused on science and technology tasks. (Definition from M2M Objectives)
Exploration Strategy	Establish the scenarios, conceptual missions, and systems needed to extend humanity's reach beyond low-Earth orbit (LEO), return to the Moon, and proceed on toward Mars and beyond
Function	Actions that an architecture would perform that are necessary to complete the desired Use Case
Global	Infrastructure and capabilities that support human and robotic operations and utilization across the subject planetary surface. (Definition from M2M Objectives)
Habitable Environment	The environment that is necessary to sustain the life of the crew and to allow the crew to perform their functions in an efficient manner.
Human Landing System - Initial Configuration	Any crewed mission to the lunar surface executed with the initial HLS configurations as defined in the HLS Broad Agency Announcement (BAA) Option A. (Effectivity for requirements unique to this configuration are noted as "HLS Initial Configuration.")

Term	Description
Human-Rating	A human-rated system accommodates human needs, effectively utilizes human capabilities, controls hazard with sufficient certainty to be considered safe for human operations, and provides, to the maximum extent practical, the capability to safely recover the crew from hazardous situations. Human-rating consists of three fundamental tenets: 1) Human-rating is the process of designing, evaluating, and assuring that the total system can safely conduct the required human missions. 2) Human-rating includes the incorporation of design features and capabilities that accommodate human interaction with the system to enhance overall safety and mission success. 3) Human-rating includes the incorporation of design features and capabilities to enable safe recovery of the crew from hazardous situations."
Hybrid Propulsion System	A vehicle consisting of two or more unique propulsion systems, each optimized for different types of maneuvers. For the purpose of this document, two hybrid systems are considered: SEP/Chem, which combines a Solar Electric Propulsion system with a Chemical stage, and NEP/Chem, which combines a Nuclear Electric Propulsion system with a Chemical stage.
Increment	The period of time between the end of one crew mission (i.e., crew splashdown) and the end of a second crew mission, including the uncrewed activities and operations that commence during this defined timeframe.
Incremental	Building compounding operational capabilities within the constraints of schedule, cost, risk, and access. (Definition from M2M Objectives)
Interoperability	The ability of two or more systems to physically interact; exchange data, information, or consumables; or share common equipment while successfully performing intended functions.
Limited Capability Mission	A mission to a polar landing site where the utilization capability of the mission is limited to the threshold capabilities of HLS and Orion, with no additional delivery or return mass available from goal capabilities or other elements. Additionally, certain missions may prioritize crew time and transportation mass to the delivery and outfitting of new elements in NRHO (e.g. Gateway elements) or the lunar surface (e.g. PR and SH). For the purposes of analysis, a 2-crew, 6.5-day sortie was assumed as a representative case. In such a mission, it is expected that a significant amount of crew time will be needed to ingress, setup, outfit, and checkout new elements being delivered to or operated for the first time in NRHO or on the lunar surface, leaving less time available for utilization activities. In addition to crew time, it is expected that the delivery and outfitting of these new elements will require a greater fraction of the overall logistics mass delivery capability, further reducing the utilization potential of the mission. Thus, this mission category represents a case in which only a threshold of utilization activities is expected to be performed.
Live	The ability to conduct activities beyond tasks on a schedule. Engage in hobbies, maintain contact with friends and family, and maintain healthy work-life balance. (Definition from M2M Objectives)
Loss of Crew	Death of or permanently debilitating injury to one or more crew members.

Term	Description
Loss of Mission	Loss of or inability to complete significant/primary mission objectives, which includes Loss of Crew. Each mission is defined with different assumptions and mission objectives. Therefore, specific mission LOM assessments are accomplished evaluating the attainment of specific mission objectives, using methods tailored to the specific mission risk drivers and each specific program but consistent with defined NASA Probabilistic Risk Assessment standards.
Maintenance	The function of keeping items or equipment in, or restoring them to, a specified operational condition. It includes servicing, test, inspection, adjustment/alignment, removal, replacement, access, assembly/disassembly, lubrication, operation, decontamination, installation, fault location, calibration, condition determination, repair, modification, overhaul, rebuilding, and reclamation. Preventative maintenance is performed before a failure occurs, whereas corrective maintenance is in response to a failure.
Mechanical Assistance	Device intended to allow the crew to transport more mass than they can hand carry while walking.
Mission	A major activity required to accomplish an Agency goal or to effectively pursue a scientific, technological, or engineering opportunity directly related to an Agency goal. Mission needs are independent of any particular system or technological solution. (Definition from M2M Objectives)
Mobility	Powered surface travel that extends the exploration range beyond what is possible for astronauts to cover on foot. Spans robotic and crewed systems and can be accomplished on and above the surface. (Definition from M2M Objectives)
Needs	A statement that drives architecture capability, is necessary to satisfy the Moon-to-Mars Objectives, and identifies a problem to be solved, but is not the solution.
Planetary Protection	Approaches used to avoid harmful contamination of solar system bodies during exploration activities, as well as avoiding possible harmful extraterrestrial contamination from material that may be returned from other solar system bodies, in compliance with Outer Space Treaty constraints.
Powered Mobility Asset	Asset that allows the crew to travel further distances than they can walk, e.g. LTV or Pressurized Rover.
Reconfiguration	If a system is required to provide a function, any time required by the crew associated with making that function available for use. Changing spaces, moving logistics to allow for use of the space for a different purpose (e.g., exercise, eating, sleeping, medical, training, working)
Routine	Recurring subject operations performed as part of a regular procedure rather than for a unique reason. (Definition from M2M Objectives)
Routine Preventative Maintenance	Planned maintenance done on a regular (daily, weekly, monthly) basis that is part of the design such as filter changes, lubrication, cleaning, etc.
Secondary Payloads	Additional cargo carried on a transportation element, currently on an adapter ring, after the primary and CPLs are accommodated, limited by the remaining transportation element resources, e.g. mass, volume, power, etc.

Term	Description
Scalability	Initial systems designed such that minimal recurring DDT&E is needed to increase the scale of a design to meet end state requirements. (Definition from M2M Objectives)
Segments	A portion of the architecture, identified by one or more notional missions or integrated use cases, illustrating the interaction, relationships, and connections of the sub-architectures through progressively increasing operational complexity and objective satisfaction.
Sol	Martian day, approximately 24 hours and 39 minutes long. For the purpose of this document, operational timekeeping on the surface of Mars will use Martian sols to align with the Martian day/night cycle.
Sortie Missions	A single crewed mission to a lunar surface location for a period of days supported solely by the lunar crewed lander. The main characteristics of the sortie mission are that crew habitation is provided by the crewed lander and the crew can perform all lunar surface activities using self-contained resources – although, pre-deployment of resources is not necessarily precluded during a sortie mission.
Sub-Architecture	A group of tightly-coupled elements, functions, and capabilities that perform together to accomplish architecture objectives.
System	The combination of elements that function together to produce the capability required to meet a need. The elements include all hardware, software, equipment, facilities, personnel, processes, and procedures needed for this purpose. (Refer to NPR 7120.5.)
To Be Determined (TBD)	Used when it is not known what value to be placed in a requirement and there is open work to determine what it should be.
To Be Resolved (TBR)	Used when a value for a requirement is presented but it is to be resolved or refined as to whether it is the right number.
Use Case	Operations that would be executed to produce the desired needs and/or characteristics
Utilization	Use of the platform, campaign and/or mission to conduct science, research, test and evaluation, public outreach, education, and industrialization. (Definition from M2M Objectives)
Utilization Mass	The total mass of all items primarily used, collected, or generated to accomplish utilization objectives in the course of the mission, including: payloads, equipment, tools, instruments, consumables, samples, containers, etc. It does not include any multi-use items that are necessary for the mission and would be carried anyway regardless of their utilization application.
Validate	Confirming that a system satisfies its intended use in the intended environment (Did we build the right system?). (Definition from M2M Objectives)
Verification (of a product)	Proof of compliance with a requirement. Verification may be determined by testing, analysis, demonstration or inspection.
Work Time	Non-personal time. Time during which the crew is in a duty status (e.g. 8-8.5 hours usually but could be 11.5 hours for an EVA day or other mission specific extension)

www.ingramcontent.com/pod-product-compliance
Lightning Source LLC
Chambersburg PA
CBHW080547220326
41599CB00032B/6389